AF865913

Air Pollution - Latest Status and Current Developments

Edited by Murat Eyvaz, Ahmed Albahnasawi and Motasem Y. D. Alazaiza

Published in London, United Kingdom

Air Pollution - Latest Status and Current Developments
http://dx.doi.org/10.5772/intechopen.104367
Edited by Murat Eyvaz, Ahmed Albahnasawi and Motasem Y. D. Alazaiza

Contributors
Diego López-Veneroni, Elizabeth Vega, Zhou Yonggang, Mingzhou Yu, Zhandong Shi, Abebaw Addisu, José L. S. Pereira, Carla Garcia, Henrique Trindade, Jorge Mendez-Astudillo, Ernesto Caetano, Karla Pereyra-Castro, Bhavya Bhargava, Mahinder Partap, Diksha Sharma, Meenakshi Thakur, Vipasha Verma, Deekshith HN, Anjali Chandel, Mehdi Farhane, Otmane Souhar, Muhammad Usman Farid, Atta Ullah, Abdul Ghafoor, Shahbaz Nasir Khan, Mazhar Iqbal, Furqan Muhayodin, Abdul Shabbir, Chaudhry Arslan, Abdul Nasir, Liliya M. Fatkhutdinova, Olesya V. Skorokhodkina, Laila I. Yapparova, Ramil R. Zalyalov, Guzel A. Timerbulatova, João Miranda Garcia, Rita Cerdeira, Luís Coelho, Murat Eyvaz, Ahmed Albahnasawi, Motasem Y. D. Alazaiza

© The Editor(s) and the Author(s) 2023
The rights of the editor(s) and the author(s) have been asserted in accordance with the Copyright, Designs and Patents Act 1988. All rights to the book as a whole are reserved by INTECHOPEN LIMITED. The book as a whole (compilation) cannot be reproduced, distributed or used for commercial or non-commercial purposes without INTECHOPEN LIMITED's written permission. Enquiries concerning the use of the book should be directed to INTECHOPEN LIMITED rights and permissions department (permissions@intechopen.com).
Violations are liable to prosecution under the governing Copyright Law.

Individual chapters of this publication are distributed under the terms of the Creative Commons Attribution 3.0 Unported License which permits commercial use, distribution and reproduction of the individual chapters, provided the original author(s) and source publication are appropriately acknowledged. If so indicated, certain images may not be included under the Creative Commons license. In such cases users will need to obtain permission from the license holder to reproduce the material. More details and guidelines concerning content reuse and adaptation can be found at http://www.intechopen.com/copyright-policy.html.

Notice
Statements and opinions expressed in the chapters are these of the individual contributors and not necessarily those of the editors or publisher. No responsibility is accepted for the accuracy of information contained in the published chapters. The publisher assumes no responsibility for any damage or injury to persons or property arising out of the use of any materials, instructions, methods or ideas contained in the book.

First published in London, United Kingdom, 2023 by IntechOpen
IntechOpen is the global imprint of INTECHOPEN LIMITED, registered in England and Wales, registration number: 11086078, 5 Princes Gate Court, London, SW7 2QJ, United Kingdom
Printed in Croatia

British Library Cataloguing-in-Publication Data
A catalogue record for this book is available from the British Library

Additional hard and PDF copies can be obtained from orders@intechopen.com

Air Pollution - Latest Status and Current Developments
Edited by Murat Eyvaz, Ahmed Albahnasawi and Motasem Y. D. Alazaiza
p. cm.

This title is part of the Environmental Sciences Book Series, Volume 9
Topic: Pollution
Series Editor: J. Kevin Summers
Topic Editors: Ismail Md. Mofizur Rahman and Zinnat Ara Begum

Print ISBN 978-1-83768-916-3
Online ISBN 978-1-83768-917-0
eBook (PDF) ISBN 978-1-83768-918-7
ISSN 2754-6713

We are IntechOpen, the world's leading publisher of Open Access books
Built by scientists, for scientists

6,600+
Open access books available

177,000+
International authors and editors

195M+
Downloads

156
Countries delivered to

Our authors are among the
Top 1%
most cited scientists

12.2%
Contributors from top 500 universities

WEB OF SCIENCE™

Selection of our books indexed in the Book Citation Index
in Web of Science™ Core Collection (BKCI)

Interested in publishing with us?
Contact book.department@intechopen.com

Numbers displayed above are based on latest data collected.
For more information visit www.intechopen.com

IntechOpen Book Series
Environmental Sciences
Volume 9

Aims and Scope of the Series

Scientists have long researched to understand the environment and man's place in it. The search for this knowledge grows in importance as rapid increases in population and economic development intensify humans' stresses on ecosystems. Fortunately, rapid increases in multiple scientific areas are advancing our understanding of environmental sciences. Breakthroughs in computing, molecular biology, ecology, and sustainability science are enhancing our ability to utilize environmental sciences to address real-world problems.

The four topics of this book series - Pollution; Environmental Resilience and Management; Ecosystems and Biodiversity; and Water Science - will address important areas of advancement in the environmental sciences. They will represent an excellent initial grouping of published works on these critical topics.

Meet the Series Editor

J. Kevin Summers is a Senior Research Ecologist at the Environmental Protection Agency's (EPA) Gulf Ecosystem Measurement and Modeling Division. He is currently working with colleagues in the Sustainable and Healthy Communities Program to develop an index of community resilience to natural hazards, an index of human well-being that can be linked to changes in the ecosystem, social and economic services, and a community sustainability tool for communities with populations under 40,000. He leads research efforts for indicator and indices development. Dr. Summers is a systems ecologist and began his career at the EPA in 1989 and has worked in various programs and capacities. This includes leading the National Coastal Assessment in collaboration with the Office of Water which culminated in the award-winning National Coastal Condition Report series (four volumes between 2001 and 2012), and which integrates water quality, sediment quality, habitat, and biological data to assess the ecosystem condition of the United States estuaries. He was acting National Program Director for Ecology for the EPA between 2004 and 2006. He has authored approximately 150 peer-reviewed journal articles, book chapters, and reports and has received many awards for technical accomplishments from the EPA and from outside of the agency. Dr. Summers holds a BA in Zoology and Psychology, an MA in Ecology, and Ph.D. in Systems Ecology/Biology.

Meet the Volume Editors

Dr. Murat Eyvaz is an associate professor in the Environmental Engineering Department, Gebze Technical University, Turkey. His research interests include applications in water and wastewater treatment facilities, electrochemical treatment processes, and filtration systems at the laboratory and pilot scale, membrane processes (forward osmosis, reverse osmosis, membrane bioreactors), membrane manufacturing methods (polymeric membranes, nanofiber membranes, electrospinning), spectrophotometric analyses (UV, atomic absorption spectrophotometry), and chromatographic analyses (gas chromatography, high-pressure liquid chromatography). He has co-authored many journal articles and conference papers and has taken part in many national projects. He serves as an editor for 50 journals and a reviewer for 120 others. He holds four patents on wastewater treatment systems.

Dr. Ahmed Albahnasawi is a research fellow in the Environmental Engineering Department, Gebze Technical University, Turkey. His graduate work focused on the investigation of the treatability of sequential anoxic-aerobic batch reactors followed by ceramic membrane for textile wastewater treatment. Dr. Albahnasawi has published three journal articles and participated in three international conferences. His research interests include the application and design of microbial fuel cells integrated with Fenton oxidation for industrial wastewater treatment/solid waste management and monitoring of organic micropollutants by both chromatographic and spectrophotometric analyses.

Dr. Motasem Y. D. Alazaiza is an assistant professor in the Civil and Environmental Engineering Department, A'Sharqiyah University, Oman. His graduate work focused on the Photographic Methods Techniques in Monitoring Non-Aqueous Phase Liquid Migration in Porous Medium. Dr. Alazaiza has published numerous journal articles and participated in numerous international conferences. His research interests include the transport of multiple fluid phases (water, oil, gases) in porous media and the mass transfer between those phases, with a particular specialization in the behavior of gases in contaminated groundwater systems. In addition, his research is applied to understand the fate and transport of non-aqueous phase liquids (NAPLs) in soil and groundwater, develop and improve remediation technologies, assess risks associated with unconventional resource development, and use a combination of laboratory experiments and numerical modeling.

Contents

Preface

Air pollution is a global environmental challenge that poses significant threats to human health and the well-being of ecosystems. *Air Pollution – Latest Status and Current Developments* is a comprehensive collection of chapters that delve into various aspects of air pollution, including its causes, effects, and potential solutions. This book provides readers with a multidimensional understanding of air pollution by exploring the latest trends and advancements in this field. Chapter 1 serves as an introductory chapter, laying the foundation for the subsequent discussions. It offers a comprehensive overview of air pollution, helping readers grasp its complex nature and understand the underlying causes that contribute to its occurrence. Furthermore, this chapter explores the wide-ranging effects of air pollution on both human health and the environment, emphasizing the urgent need for effective solutions. Chapter 2 focuses on the relationship between air pollution and clean energy. It provides valuable insights into the latest trends and future perspectives in utilizing clean energy sources to mitigate air pollution. The chapter explores innovative technologies and strategies that hold promise in reducing pollutant emissions and creating a sustainable future. Chapter 3 explores the concept of phytoremediation to combat air pollutants. The chapter highlights the latest advancements in this field, shedding light on the potential of plants to remove contaminants from the air. The discussions encompass various phytoremediation techniques, their effectiveness, and their applicability in different contexts. Chapter 4 delves into the intricate relationship between fine particles in the ambient air and their potential implications as a risk factor for bronchial asthma in adults, offering a deeper understanding of the complex interplay between air quality and public health.

The following chapters continue to unveil intriguing aspects of air pollution. Chapter 5 introduces a novel development in three-dimensional analytical solutions for air pollution dispersion modeling, presenting advanced methodologies to improve our understanding of pollutant dispersion and its spatial distribution. Chapter 6 provides a comprehensive review of particle removal through thermophoretic deposition, shedding light on this intriguing phenomenon and its implications for air quality control. Indoor air pollution, a critical yet often overlooked issue, takes center stage in Chapter 7. The chapter explores the sources, effects, and control measures related to indoor air pollutants, emphasizing the importance of ensuring clean and healthy indoor environments. Chapter 8 focuses on measures to control airborne pollutants specifically in broiler housing. It presents a thorough review of various strategies and technologies aimed at mitigating the adverse effects of air pollution on poultry farms, highlighting the significance of maintaining optimal air quality in these settings. Chapter 9 takes a closer look at the air quality in Mexico City following public policy interventions. It examines the effectiveness of these interventions in improving air quality and analyzes the associated challenges and opportunities. Chapter 10 introduces the use of stable isotopes as a tool to identify carbon and nitrogen sources in Mexico City's PM2.5 during the dry season. This chapter presents innovative techniques for source apportionment, facilitating a better understanding of pollution origins and aiding in the development of targeted mitigation strategies. Lastly, Chapter 11 investigates the impact of air quality on children's health, specifically in Barreiro. By analyzing temporal lag, this chapter offers insights into the delayed effects of air pollution exposure on the well-being of children, highlighting the importance of proactive measures to safeguard their health.

Air Pollution – Latest Status and Current Developments provides a comprehensive exploration of various dimensions of air pollution. Each chapter offers valuable insights into the causes, effects, and potential solutions, presenting the latest advancements in the field. We hope that this book will serve as a valuable resource for researchers, policymakers, and anyone interested in understanding and addressing the complex issue of air pollution. The diverse range of topics covered in this book displays the interdisciplinary nature of air pollution research, highlighting the need for collaborative efforts across various scientific disciplines. By bringing together experts from different fields, this book aims to foster a holistic understanding of air pollution and promote the development of effective strategies to mitigate its adverse effects. It is important to acknowledge that the fight against air pollution requires a collective effort. Governments, policymakers, industry stakeholders, and individuals all have a crucial role to play in implementing sustainable practices and policies that reduce pollutant emissions and promote cleaner air. This book serves as a platform for knowledge sharing, providing evidence-based insights that can inform decision-making processes and inspire action. While this book captures the latest status and current developments in air pollution research, it is essential to recognize that the field is ever-evolving. New challenges and opportunities will continue to arise as our understanding of air pollution deepens and new technologies and strategies emerge. Therefore, we hope that this book will inspire further research, innovation, and collaboration in the pursuit of cleaner and healthier air for all.

We would like to express our sincere gratitude to the authors who have contributed their valuable insights and expertise to this book. Their dedication and commitment to advancing the field of air pollution research have made this publication possible. We would also like to extend our thanks to the reviewers and editorial team who have worked diligently to ensure the quality and integrity of the content. Finally, we would like to thank the readers for their interest in this book. We hope that the chapters presented within these pages will serve as a catalyst for meaningful discussions and inspire future endeavors to combat air pollution. Together, let us strive towards a cleaner and more sustainable future, where the air we breathe is free from harmful pollutants.

Murat Eyvaz
Associate Professor,
Department of Environmental Engineering,
Gebze Technical University,
Kocaeli, Turkey

Dr. Ahmed Albahnasawi
Department of Environmental Engineering,
Gebze Technical University,
Kocaeli, Turkey

Motasem Y.D. Alazaiza
Department of Civil and Environmental Engineering,
A'Sharqiyah University,
Ibra, Oman

Chapter 1

Introductory Chapter: Air Pollution – Understanding Its Causes, Effects, and Solutions

Murat Eyvaz, Ahmed Albahnasawi and Motasem Y.D. Alazaiza

1. Introduction

Air pollution is a significant environmental and public health issue that affects millions of people worldwide. It is caused by a variety of human activities, including industrial processes, transportation, and energy production, among others. The problem is particularly severe in urban areas, where population density and economic activity contribute to high levels of pollution. Air pollution can have significant health impacts, including respiratory diseases, cardiovascular diseases, and cancer, among others. In addition to its impact on human health, air pollution can also have adverse effects on the environment, including on plant and animal species and ecosystems.

Air pollution levels vary greatly between regions and countries and are influenced by a range of factors such as climate, topography, and population density. In urban areas, air pollution is often higher due to the concentration of anthropogenic sources, such as traffic, industry, and power generation. For example, a study conducted in Beijing, China, found that the city's air pollution was primarily caused by the burning of fossil fuels, including coal, oil, and natural gas [1].

Air pollution is also a major environmental concern in developing countries, where industrial activities and transportation infrastructure are expanding rapidly. In India, for instance, air pollution is estimated to cause over one million premature deaths each year [2]. The country has implemented several measures to address air pollution, including the National Clean Air Program, which aims to reduce particulate matter concentrations by 20–30% by 2024 [3].

This book aims to provide an overview of the current state of air pollution and the latest developments in this field. It covers a range of topics, including the sources and types of air pollutants, their effects on human health and the environment, and the policies and technologies aimed at reducing emissions and improving air quality. It is intended for students, researchers, policymakers, and anyone interested in understanding and addressing this critical environmental issue.

2. Sources of air pollution

There are many sources of air pollution, including both human-made and natural sources. Human-made sources of air pollution include industrial activities, transportation, energy production, and agricultural practices. These sources release a range of pollutants, including particulate matter (PM), nitrogen oxides (NOx), sulfur dioxide (SO2), volatile organic compounds (VOCs), and carbon monoxide (CO), among others. Natural sources of air pollution include wildfires, dust storms, and volcanic eruptions, among others.

3. Health impacts of air pollution

Air pollution can have significant health impacts, particularly for vulnerable populations such as children, the elderly, and individuals with pre-existing health conditions. The World Health Organization (WHO) estimates that air pollution is responsible for approximately seven million premature deaths annually worldwide [4]. Exposure to air pollution can lead to a range of health problems, including respiratory diseases such as asthma and chronic obstructive pulmonary disease (COPD), cardiovascular diseases, and cancer, among others. Long-term exposure to air pollution has also been linked to cognitive decline and neurological disorders [5].

Recent research has highlighted the health impacts of air pollution, particularly the impact of PM. A study published in The Lancet Planetary Health found that exposure to PM2.5 (fine particulate matter with a diameter of less than 2.5 micrometers) is responsible for approximately 500,000 premature deaths annually in Europe [6]. Another study published in Environmental Research estimated that long-term exposure to PM2.5 is responsible for 6.7 million premature deaths annually worldwide [7]. Another studies have highlighted the impact of air pollution on cognitive function, with exposure to high levels of air pollution linked to decreased cognitive performance and an increased risk of dementia [8, 9]. Other studies have focused on the impact of air pollution on plant and animal life, with findings showing that air pollution can have significant negative effects on ecosystems and biodiversity [10].

4. Environmental impacts of air pollution

Air pollution can also have adverse effects on the environment, including on plant and animal species and ecosystems. Acid rain, for example, is a type of air pollution that can have significant impacts on forests, lakes, and rivers. Acid rain occurs when sulfur dioxide and nitrogen oxides are released into the atmosphere and react with water, oxygen, and other chemicals to form acidic compounds. These compounds can then fall to the ground as acid rain, damaging forests, lakes, and rivers and harming plant and animal species.

5. Climate change and air pollution

Air pollution is also a significant contributor to climate change. Greenhouse gases, such as carbon dioxide (CO_2), trap heat in the atmosphere, causing global temperatures to rise. Human activities, including the burning of fossil fuels, transportation, and industrial processes, contribute to the release of greenhouse gases and the warming of the planet. Climate change can have significant environmental and social impacts, including rising sea levels, increased frequency and intensity of natural disasters, and food and water scarcity, among others.

6. Current developments in air pollution control

Governments, businesses, and individuals can all play a role in reducing air pollution through policies, investments, and behavior change. Many countries have implemented policies to reduce air pollution, including regulations on industrial emissions, cleaner energy standards, and the promotion of public transportation and active transportation options. One such initiative is the Paris Agreement, which aims to limit global warming to below 2°C above pre-industrial levels by reducing greenhouse gas emissions [11]. Another initiative is the World Health Organization's Global Ambient Air Quality Database, which provides information on air quality levels in cities and countries worldwide [4].

Technological developments and innovations are also contributing to the fight against air pollution. For example, electric vehicles and renewable energy sources are becoming increasingly popular and affordable, reducing the need for fossil fuels and decreasing emissions. Advances in monitoring technology, such as air quality sensors and satellite imagery, are also providing more accurate and real-time data on air pollution levels.

7. Challenges and future directions

Despite these promising developments, there are still significant challenges to addressing air pollution. In many parts of the world, particularly in developing countries, air pollution levels are still high, and policies and regulations may not be adequately enforced. The problem of air pollution is also complex, and solutions may require significant changes in infrastructure, behavior, and policy.

There is a growing recognition of the need for a coordinated, global effort to address air pollution. The United Nations Sustainable Development Goals include a target to substantially reduce the number of deaths and illnesses from air pollution by 2030. Achieving this target will require a range of strategies, including investments in clean energy, improvements in public transportation, and more effective regulation of industrial activities. Moreover, cleaner technologies, such as electric vehicles and renewable energy sources, as well as the adoption of policies aimed at reducing emissions from existing sources [12] were proposed. In addition, there is a growing interest in the use of green infrastructure, such as urban forests and green roofs, to improve air quality in urban areas [13].

8. Conclusion

Air pollution is a significant environmental and public health issue that affects millions of people worldwide. It is caused by a variety of human activities, including industrial processes, transportation, and energy production. Air pollution can have significant health impacts, particularly for vulnerable populations such as children, the elderly, and individuals with pre-existing health conditions. In addition to its impact on human health, air pollution can also have adverse effects on the environment, including on plant and animal species and ecosystems. Technological developments and innovations are contributing to the fight against air pollution, but more work is needed to address the problem. Governments, businesses, and individuals can all play a role in reducing air pollution through policies, investments, and behavior change.

Author details

Murat Eyvaz[1*], Ahmed Albahnasawi[1] and Motasem Y.D. Alazaiza[2]

1 Department of Environmental Engineering, Gebze Technical University, Kocaeli, Turkey

2 Department of Civil and Environmental Engineering, A'Sharqiyah University, Ibra, Oman

*Address all correspondence to: meyvaz@gtu.edu.tr

IntechOpen

© 2023 The Author(s). Licensee IntechOpen. This chapter is distributed under the terms of the Creative Commons Attribution License (http://creativecommons.org/licenses/by/3.0), which permits unrestricted use, distribution, and reproduction in any medium, provided the original work is properly cited.

References

[1] Wang L, Zhang F, Pilot E, Yu J, Nie C, Holdaway J, et al. Taking action on air pollution control in the Beijing-Tianjin-Hebei (BTH) region: Progress, challenges and opportunities. International Journal of Environmental Research and Public Health. 2018;**15**(2):306. DOI: 10.3390/ijerph15020306

[2] Landrigan PJ, Fuller R, Acosta NJR, Adeyi O, Arnold R, Basu NN, et al. The lancet commission on pollution and health. Lancet (London, England). 2018;**391**(10119):462-512. DOI: 10.1016/S0140-6736(17)32345-0

[3] Ministry of Environment, Forest and Climate Change. National Clean Air Programme (NCAP). Government of India. New Delhi, India. Available from: https://moef.gov.in/wp-content/uploads/2019/05/NCAP_Report.pdf. 2019 Access date: 05/04/2023

[4] World Health Organization. Ambient (outdoor) pollution. 2022. Retrieved from: https://www.who.int/news-room/fact-sheets/detail/ambient-(outdoor)-air-quality-and-health [Access date: 03/04/2023]

[5] Jiang XQ, Mei XD, Feng D. Air pollution and chronic airway diseases: What should people know and do?. Journal of Thoracic Disease. 2016;**8**(1):E31-E40. DOI: 10.3978/j.issn.2072-1439.2015.11.50

[6] Lelieveld J, Klingmüller K, Pozzer A, Pöschl U, Fnais M, Daiber A, et al. Cardiovascular disease burden from ambient air pollution in Europe reassessed using novel hazard ratio functions. European Heart Journal. 2019;**40**(20):1590-1596. DOI: 10.1093/eurheartj/ehz135

[7] Cohen AJ, Brauer M, Burnett R, Anderson HR, Frostad J, Estep K, et al. Estimates and 25-year trends of the global burden of disease attributable to ambient air pollution: An analysis of data from the global burden of diseases study 2015. Lancet (London, England). 2017;**389**(10082):1907-1918. DOI: 10.1016/S0140-6736(17)30505-6

[8] Zhang X, Chen X, Zhang X. The impact of exposure to air pollution on cognitive performance. Proceedings of the National Academy of Sciences of the United States of America. 2018;**115**(37):9193-9197. DOI: 10.1073/pnas.1809474115

[9] Cai R, Zhang Y, Simmering JE, Schultz JL, Li Y, Fernandez-Carasa I, et al. Enhancing glycolysis attenuates Parkinson's disease progression in models and clinical databases. Journal of Clinical Investigation. 2019;**129**(10):4539-4549

[10] Lovett GM, Tear TH, Evers DC, Findlay SE, Cosby BJ, Dunscomb JK, et al. Effects of air pollution on ecosystems and biological diversity in the eastern United States. Annals of the New York Academy of Sciences. 2009;**1162**:99-135. DOI: 10.1111/j.1749-6632.2009.04153.x

[11] United Nations. Paris agreement. 2015. Available from: https://unfccc.int/sites/default/files/english_paris_agreement.pdf [Access date: 03/04/2023]

[12] Sims R, Schaeffer R, Creutzig F, Cruz-Núñez X, D'Agosto M, Dimitriu D, et al. Transport. In: Edenhofer O, Pichs-Madruga R, Sokona Y, Farahani E, Kadner S, Seyboth K, editors. Climate Change 2014: Mitigation of Climate Change. Contribution of Working Group III to the Fifth Assessment Report of the

Intergovern-Mental Panel on Climate Change. Cambridge, United Kingdom and New York, NY, USA: Cambridge University Press; 2014

[13] Nowak DJ, Hirabayashi S, Bodine A, Hoehn R. Modeled PM2.5 removal by trees in ten U.S. cities and associated health effects. Environmental pollution. 2013;**178**:395-402. DOI: 10.1016/j.envpol.2013.03.050

Chapter 2

Air Pollution and Clean Energy: Latest Trends and Future Perspectives

Muhammad Usman Farid, Atta Ullah, Abdul Ghafoor, Shahbaz Nasir Khan, Mazhar Iqbal, Furqan Muhayodin, Abdul Shabbir, Chaudhry Arslan and Abdul Nasir

Abstract

Energy and the environment are among the top global issues of this era. Environmental degradation specifically due to consumption of fossil fuels in conventional energy generation systems has become a critical challenge for the whole world. With the introduction of advance industrial processes and operations, the air quality deterioration has also become very complex. There is a dire need to replace the conventional energy systems with alternative energy resources for reducing air pollutants. Renewable energy systems generate clean energy with less environmental footprints. This chapter will highlight the latest trends and future strategies in clean and renewable energy supply systems to mitigate air pollution for environmental sustainability.

Keywords: air quality, clean energy, fossil fuels, energy and environment, Air pollution: Sources types and classification

1. Introduction

Air pollution has become a major health hazard to the millions of people around the globe. It has been linked to asthma, heart disease, and other serious health problems. Air pollution triggers asthma, which is a long-term disease of the lungs that affects the airways and makes it hard to breathe. Tiny particles in the air get deep into the lungs and cause inflammation, which can lead to an asthma attack. Cardiovascular diseases such as heart attacks and strokes have also been linked to air pollution. Nitrogen oxides and sulfur dioxide are also causes of oxidative stress, which can lead to heart problems. Various studies have shown that the exposure to air pollution increases the probability of early decay and death among the living organisms. This is especially true for the elderly and people who already have health problems. Adult's lung function is adversely correlated with PM10, nitrogen dioxide, and sulfur dioxide, all of which have been linked to bronchitis symptoms in studies encompassing eight different communities. Research into the effects of air pollution has also revealed a decline in human lifespan [1].

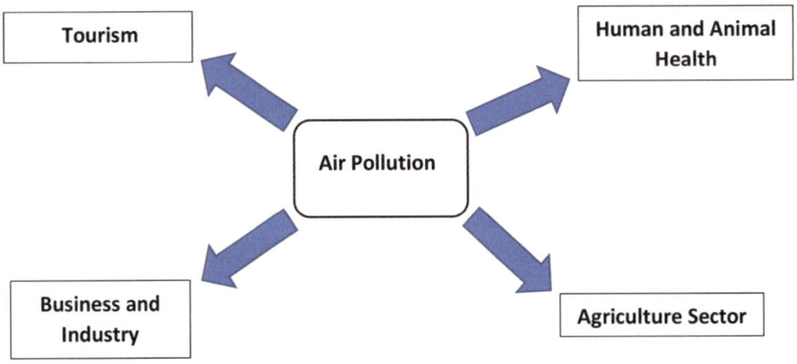

Figure 1.
Effects of air pollution on various sectors.

Air pollution not only harm people's health, but also affects the environment in various ways (**Figure 1**). Here are some primary ways that pollution in the air can affect nature. Air pollutants in form of acid rain have detrimental effects on crops and forests growth [2]. Furthermore, the damage of leaves hinders crop yields [3]. This issue contributes to food insecurity and disturbs plant ecosystems. For instance, acid rain makes lakes and streams more acidic, and it is hard for certain species to survive. Similarly, carbon dioxide, methane, and other greenhouse gases are examples of air pollutants that contribute significantly to global warming and climate change [4]. This results in glaciers melting, rising sea levels, and stronger weather events. The poor air quality causes smog and haze, which makes it difficult for the people to enjoy the outdoor activities [5]. Moreover, air pollution can also have a significant negative effect on various businesses, tourism industry, and ultimately the economy of the region [6].

In order to mitigate air pollution, it is very important to understand the origin, sources, and classification of air pollutants. Moreover, the life cycle of the air pollutant can provide a better insight to manage the potential sources causing air pollution in a system. The major contributor in deterioration of air quality is the energy sector; hence, it is also important to study air pollution in connection with the energy generation systems for sustainability of environment as well as communities. Keeping in view such factors, this chapter summarizes a thorough overview of air pollution, including its causes, types, and classification as well as the use of clean energy resources for minimizing the discharge of pollutants and contaminants in the air. Different factors associated with air pollution such as its effects on health, environment, climate, and the most recent trends in air pollution are also discussed.

1.1 Overview of air pollution

Pollution is addition of any foreign matter, which changes the purity of any system. Hence, air pollution can be defined as "it is the addition of any foreign and undesirable matter generated from physical, chemical, or biological sources which alters the natural quality of ambient air." Having complex mechanisms composed of both primary and secondary pollutants formulations, it is very difficult to exactly determine the air pollution. It might be argued that the use of fossil fuels and primary/finite energy resources is one of the main causes of air pollution. All anthropogenic effluents and emissions into the air are considered as air pollution due to their impact on atmospheric chemistry. Using this definition, the rise in atmospheric levels of greenhouse gases CO_2, CH_4, and N_2O can be considered as air pollution, even though these levels are not yet known to be harmful to humans or ecosystems. Air pollution can narrowly be defined as harmful chemicals released into the atmosphere by humans. The term "harmful" can refer to a variety of negative outcomes, that is, damage to manmade or naturally occurring inanimate structures, and a decrease in visibility. A chemical may have no immediate negative effects before being released into the environment [7, 8]. Hence, it is important to classify the discrete sources of air pollution for its proper mitigation and minimization.

1.2 Sources of air pollution

Pollutants in the air are divided into groups based on their origin. Based on such classification system, it becomes easy for the environmentalists to make policy decisions for specifying the release and control of air pollutant in the air.

The sources of air pollution are classified into following three major groups (**Table 1**):

1.2.1 Natural sources of air pollution

This category includes the addition of undesirable elements in the ambient air particularly, due to natural incidents occurring in the earth's atmosphere. Such elements that pollute the air include sand or dust particles, forest fires, pollen, volcanic eruptions, SMOG, release of gases from organic matters, etc. These pollutants are liberated as by-products from the cycles or incidents occurring naturally.

Sandstorms are one of the most common sources, which deteriorate the air quality. The immediate effect of sandstorm is to worsen the clarity of air and reducing the visibility and ultimately causing road accidents and difficulties in transportation. Dust particles are more likely to absorb toxic gases, which can cause severe reactions to form secondary pollutants. This also becomes a source of toxic organic compounds

Sources of air pollution	Natural sources	• Sand/dust storms • Forest fires • Volcanic eruptions • Pollen • Break down of OM
	Industrial sources	• Emissions from fossil fuels • Melting process • Processing and treatment plants • Crushing and grinding of stones • Oil and gas refineries • Fertilizers and pesticides
	Municipal sources	• Landfill gas • Sewerage • Street cleanings/dust removal • Household cooking

Table 1.
Sources of air pollution [9–19].

that make aerosols in the urban environment more toxic and biologically reactive. Dust particles act as media to transport bacteria. Studies have shown that microbe-rich dust particles can make allergic inflammation worst [20].

Another source for air pollution is forest fires, which occur as a result of natural accidents like lightning or temperature increase in the certain region that has sufficient biomass to burn up. Open fire at a massive scale in the forest releases a lot of smoke and small particles into the air. Pollutants such as oxides of carbon, that is, carbon dioxide (CO_2) and carbon monoxide (CO), oxides of nitrogen and sulfur in the form of NOx and SOx, volatile organic compounds (VOCs), as well as fine ash and particulate matter are produced from forest fire. These pollutants can lead to health problems like asthma and lung diseases [21].

Volcanic eruptions are one of the most natural and powerful sources of air pollution. Volcano eruption produces aerosol clouds that can travel thousands of miles, blocking the sun and causing health problems, crop damage, and other types of environmental damage. Volcanic eruptions can have far-reaching and terrible effects on the environment in the long run [22].

Similarly, breaking down of organic matter naturally is a major source of air pollution. As organic matter is destructed by the microorganisms in an uncontrolled manner, it gives off gases called volatile organic compounds (VOCs) and other air pollutants. These pollutants can be harmful and can contribute in the formation of smog and ozone, both of which are hazardous to health. Furthermore, such process of organic matter decomposition can also release methane (a strong greenhouse gas), which significantly contributes to climate change and global warming around the world [23].

1.2.2 Industrial or manmade sources of air pollution

Industrial sources of air pollution include all kinds of particulates and gases, which are generated by industrial activities. Stack emissions, that is, release of carbon monoxide, oxides of sulfur dioxide, and oxides of nitrogen due to the burning of fossil fuels, are the most significant elements under this category. Uncontrolled burning of fossil fuels in the energy generation systems at the industrial power units results in complete burning, which liberates toxic pollutants. These pollutants can cause a wide range of health problems, such as cancer and lung diseases. Moreover, such elements are also responsible for the acid rain that damages the local environment and the ecosystem [9].

Metal melting, preparing chemicals, treatment of fluids, and production of cement are all examples of industrial processes that release pollutants into the air [10]. Similarly, farming activities, that is, the use of chemicals in the form of pesticides, fertilizers as well as transportation activities are also the sources of air pollution. Loose and uncontrolled burning of agricultural residues and leftovers has also a major share in air pollution by adding combustion gases directly to the environment leading to the formation of smog and other secondary pollutants [11, 12].

The extraction of minerals and oil and gas refineries are the sources to add sulfur dioxide, VOCs, oxides of nitrogen, and heavy metals into the air [13, 14].

1.2.3 Municipal and domestic sources of air pollution

The municipal and domestic sources of air pollution include methane and other toxic gases produced in the landfills, which contribute to deterioration of air quality in the habitat [15]. Uncontrolled sewage gas is released in the municipal sewerage

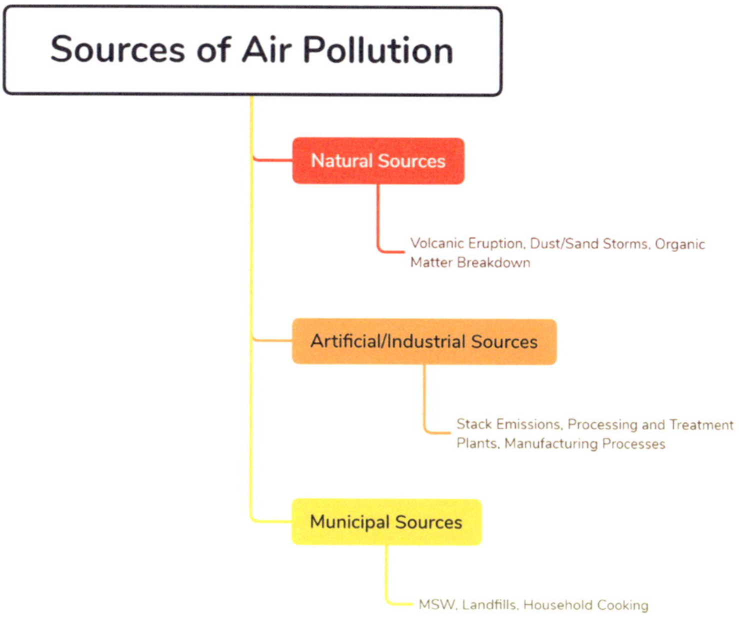

Figure 2.
Sources of air pollution.

systems, which may cause addition of potential pollutants and toxic fumes in the air [16]. Household activities such as cooking are also a source of pollutants such as particulate matter and volatile organic compounds as shown in **Figure 2** [17–19].

The above sources generically include the major contributors in the deterioration of air quality. However, air pollution dates back to the industrial revolution in the nineteenth century. At that time, people burned coal for energy and transportation, which made the air in cities much polluted. The Great Smog of 1952 in London, England, was one of the first documented cases of air pollution. Thousands of people were affected, which led to the formulation of the Clean Air Act in 1956 [24]. In the 1960s and 1970s, after being more aware about the possible effects of poor air quality and contaminants on health and ecosystem, further strict rules were also structured such as the Clean Air Act Amendment, which was passed in 1963. The formulation of Environmental Protection Agency (EPA) in the United States in 1970 also played an important role to control air pollution (Environmental Protection Agency, 2023). In the last few decades, air pollution has become a global issue. The rapid growth of industry and transportation has affected the air quality badly.

2. Potential air pollutants

As described earlier, the air pollutants are generated from the various sources and undergo a series of complex mechanisms that ultimately affect the natural environment. However, the major identified pollutants generated from the above sources are discussed as below:

2.1 Particulate matter (PM)

Particulate matter (PM) is one of the basic air pollutants, which is formed when different pollutants react chemically. The particulate matter has different particle sizes ranging from 10 μm to very fine particles having particle size of 0.5 μm or less. Such particles are made up of tiny drops of liquid or solid that can be breathed in by human and cause serious health issues such as lung infections and contamination of even bloodstream [25–28]. Both short-term and long-term effects are caused by PMs. Several epidemiological studies have been conducted to investigate such effects. Such effects include asthma, pneumonia, respiratory, and heart diseases [29, 30].

2.2 Ozone (O_3)

Ozone or ground-level ozone is formed due to reaction of flue gases (particularly VOCs and other hydrocarbon) released during the incomplete combustion of fossil fuels. This is a highly reactive gas that can cause coughing, wheezing as well as difficulty on enough air. This is considered as 52% more stronger oxidant as compared to chlorine [31–35].

2.3 Oxides of nitrogen (NOx) and sulfur (sox)

Oxides of nitrogen and sulfur are formed during uncontrolled and incomplete combustion of fossil fuels. Oxides of nitrogen include NO (nitrogen monoxide), NO_2 (nitrogen dioxide), and N_2O (di-nitrogen oxide). These gases through different chemical reactions are responsible for smog formation as well as acid rain. Similarly, oxides of sulfur are formed due to the impurities found in fuel being burnt in the firing system. It is very important to prevent these gases before they enter in the environment [36–38].

3. Air pollution and climate change

Air pollution is directly linked to the climate change and global warming. Some of the effects include rising sea levels, heatwaves that happen more often, droughts, and extreme weather events such as hurricanes, floods, and wildfires. The loss of biodiversity and changes in agricultural production are also the negative outcomes of the climate change. Almost all kinds of air pollutants are directly or indirectly involved in global warming resulting in uncertain climatic conditions. For instance, emissions from vehicles, industries, and domestic activities including combustion by-products comprising of carbon mono oxide, carbon dioxide, nitrogen oxides as well as char particles (in the form of particulate matter). Such particulate matters absorb or scatter radiation with greater capacity as compared to air and hence impart a direct impact on climate change, which may harm human, animal as well as plant health. Similarly, methane is another pollutant that contributes to climate change as its global warming potential is much higher than carbon dioxide. Possible options to cope with climatic issues and ambient air quality include an improvement in the environmentally efficient energy generation systems, less thermal losses, and use of clean energy resources to reduce emission levels around the globe [39–42]. Controlling air pollutant can have a direct and positive impact on the climate change and improves quality of environment. For this, it is very important to revise the policies for energy generation, which can minimize the consumption of fossil fuels.

4. Clean energy: introduction and overview

Energy is one of the main sectors, which contributes a major share in the environmental degradation, particularly in air phase of the environment. Conventionally, fossil fuels are used for meeting the environment requirements of industrial, commercial, transportation, domestic as well as agricultural operations and activities. Fossil fuels are excavated from the earth and after going through a series of refining processes are consumed in the energy generation systems. Burning of such fuels in the firing system acts as a baseline in the formation of primary air pollutants. Hence, technical aspects of the fossil fuel burning in the firing system are extremely important to minimize the environmental impacts. Such measures include use of refined fuels, design of combustion systems, firing methods, air to fuel ratios, filtration and scrubbing systems for smoke in the stacks as well as quality of fuel. Despite several measures, there is still likely chance of release of harmful gases from the burning of fossil fuels, which may seriously affect the air environment. Moreover, rapid depletion of fossil fuels results in high energy prices around the world. In this case, this causes both high economic costs and environmental costs. It is dire need of the time to introduce alternative energy generation ways, which can replace the need of fossil fuels (**Figure 3**) [43].

The economic, social as well as environmental constraints caused by the use of fossil fuels in energy generations system have pushed the community to use the renewable energy resources for meeting the energy requirements. Having no emissions and effluents, renewable energy resources are considered clean energy resources. The renewable energy resources, that is, clean energy resources are categorized in different groups such as Solar Energy (SE), Wind Energy (WE), Geothermal Energy (GTE), Bioenergy (BE), and Hydro Energy (HE) as shown in **Figure 4**.

Having a natural cycle in the earth's atmosphere, all kinds of clean energy resources (**Figure 4**) have zero or minimal impacts on the environment and hence

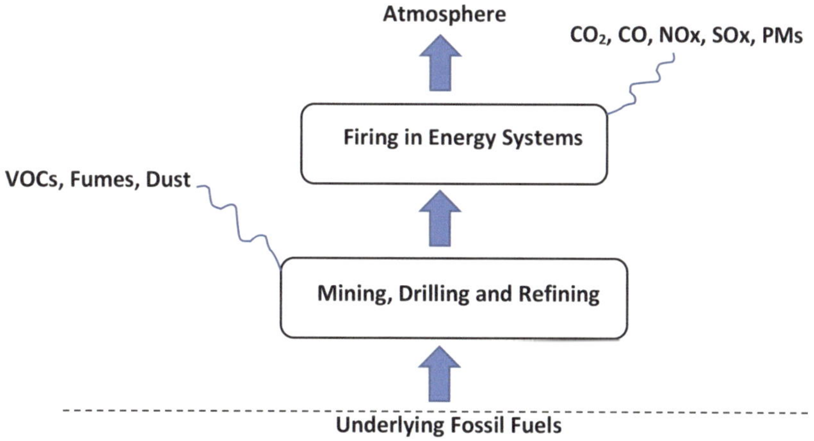

Figure 3.
Sources of emissions from fossil fuels into the atmosphere.

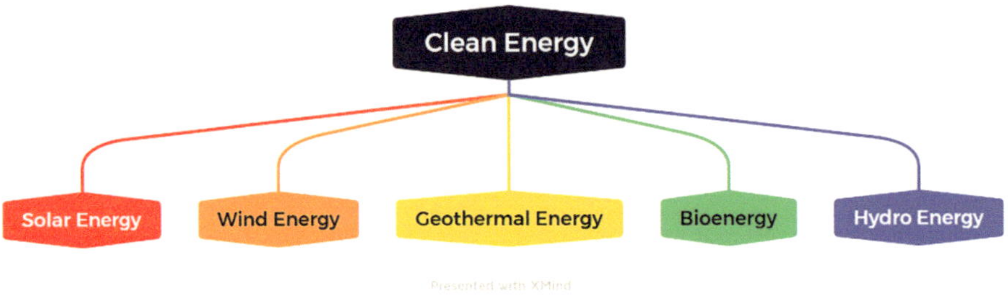

Figure 4.
Types of clean energy.

contribute to mitigate the air pollution as well as greenhouse gases. This creates more balance in the earth's environment contrary to the fossil fuels that disturb the natural balance and cause natural hazards. Further, there are several benefits to invest in such alternative energy generation techniques including lowering the energy costs, fulfillment of energy gaps due to rapid population growth, minimal environmental impacts, improved public health sustainable, creation of employment and business opportunities, and hence, development of societies and communities [44]. The modern technological advancements and innovations have made such resources more efficient and cost-effective and easier to use in place of conventional fuels and primary energy systems [45].

Different incentives and subsidies are being provided around the world for the promotion and adoption of clean energy resources in domestic, commercial as well as industrial applications around the world. However, there is a need of more friendly policies for the encouragement of such technologies so that social implications can be avoided for the successful adoption of clean energy [46].

It is important to study and understand the basis and technical aspects of clean energy resources for successful adoption. The major clean energy resources and their systematic flow of energy generation in line with its connection to mitigate air pollution are discussed as below.

4.1 Solar energy as clean energy resource

Sun is the most prominent and basic source of energy for the Earth. This is also considered to be the first-stage energy source for all kinds of clean energy resources as the energy flows in the form of solar radiation from the sun's surface toward earth's environment and is absorbed or captured by various ways. This is considered as renewable resource as a continuous flux of solar energy (solar rays) rays is received by the earth. A portion of these rays is absorbed by the clouds, dust particles, or moisture/gas molecules, another portion is reflected back to the environment, and a significant part is absorbed by the earth surface [47, 48]. If a device is placed in the path of solar rays, the energy possessed in the solar radiation can be captured to process further to the desired form of energy (**Figure 5**).

The energy flux coming from the solar source can be captured for two purposes, that is, 1—Solar Photovoltaic (PV) generation and 2—Solar Thermal Energy (**Figure 5**). The former is based on a simple P-N junction diode cell to capture photons present in the sunlight to produce a flow of electrons, which is used as an electrical energy input for various applications. Another application of solar energy is to capture and concentrate the solar radiation to produce a high-density flux of radiation resulting

Figure 5.
Outputs of solar energy [49].

in a significant temperature output, which can be used in various thermal energy applications such as water heating, drying of agricultural products, cooking, and processing of chemicals and solutions. The device used for such purpose is termed as Solar Collector and Contractors. In solar collectors, the solar radiation is captured and absorbed by the black body present in the solar collector, while in case of solar concentrators the solar radiation is contracted with the help of mirrors and focused lenses to produce a high temperature. By using both techniques, the energy requirement in both forms, that is, electricity and thermal energy can be fulfilled, which is conventionally provided by the burning of fossil fuels and other primary energy resources. A multiscale installation of such system can provide sufficient energy for town. There are certain technical challenges such as solar fluctuations, day of the time, and climate and weather conditions, which hinder the proper availability of solar rays. However, these aspects can be managed by using storage systems as well as installation of auxiliary energy system, which can be used during the time of no or less solar energy. The use of solar energy is helpful to reduce dependence on fossil fuels to meet both electricity and thermal energy for industrial and domestic applications [49].

Adoption of solar energy also provides a promising solution for the transport sector, which is one of the significant contributors for air pollution and environmental degradation. Electric vehicles (EVs) powered with solar energy are one of the alternative solutions to the conventional transportation vehicles being run by firing the fossil fuels like gasoline, diesel. The use of EVs does not release any harmful pollutants into the air. This is particularly important for the big and populated cities where transportation is major cause of respiratory and heart diseases. Availability of solar PV powered recharging stations can further help to further disseminate this environmentally friendly technology [50].

Solar energy could be combined with other clean energy sources, like wind, bioenergy, and hydroelectric power, to make a more reliable and efficient energy system. There are still certain challenges and barriers that make it hard for solar energy to be used at a large scale. However, this also highlights a lot of opportunities and future directions for introduction of modern tools and systems, which can eliminate the technical problems and promote further sustainability. In this regard, life cycle assessment, energy balances as well as energy routing of solar PV cells and collectors are also needed for the technological advancement of this environmental technology.

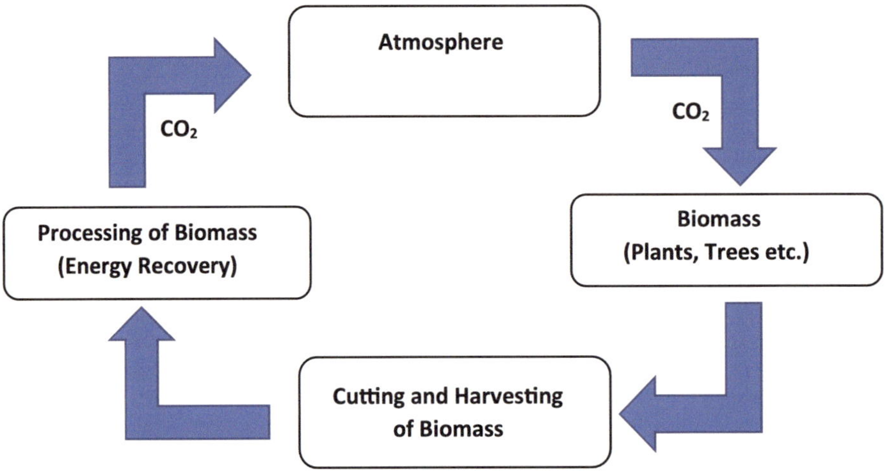

Figure 6.
Closed biomass cycle (zero net emissions) [51].

4.2 Bioenergy as clean energy resource

Bioenergy is another type of clean energy resources, which is also considered to be an alternative to many fossil fuels and conventional energy resources. Bioenergy is based on the energy extraction from biomass materials, i.e., organic matter generated from living bodies such as plants and animals. Use of biomass materials for the production of energy is considered as net zero emission technique on the basis of natural cycle of plants and organic matter. Under this concept, the same amount of carbon dioxide as well as temperature is returned to the natural environment, which was taken during the growth of biomass matter such as plants (**Figure 6**). This makes it clean source of energy as compared to the fossil fuels that are taken up from the earth or mines and after burning an additional volume of greenhouses gases is added in the atmosphere [51–53].

Biomass materials are generated as by-products during the decay of forest trees, crop production as well as raring of livestock animals. Such materials consist of significant energy contents in the form of calorific value or heating value (MJ/k), which can be recovered to meet the energy requirements. This recovery can be executed in various ways (**Table 2**).

In simple and basic energy recovery techniques, the raw biomass material is cut, chopped, and physically processed to produce a more refined form of organic matter, which can be used in other processing techniques with higher efficiency. For example, biomass materials are crushed and ground into fine particles, which are then densified with high pressure and are allowed to pass through the dyes of known size resulting in small fuel structures in the form of pellets or briquettes termed as Refused Derived Fuel. In case of oily biomass and seeds, the oil extraction technique is used. Similarly, in Thermo-Chemical Processing, biomass materials are treated and destructed under controlled temperature conditions to produce high temperature, pyrolysis oil, or more refined gaseous fuels, which can be used in place of conventional fuels. In Biological Processing, the breakdown of biomass materials occurs biologically, that is, by the microorganisms to produce secondary fuels like biogas, biodiesel (**Table 2**). Being the part of natural cycle, biomass can be used to replace

Sr. No.	Processing	Technique used	Output
1	Physical processing	• Cutting and chopping • Densification • Pelleting • Oil extraction	• Refused derived fuel (RDF) • Biofuel pellets • Oil
2	Thermo-chemical processing	• Combustion/incineration • Pyrolysis/electrolysis • Gasification • Transesterification	• Thermal energy • Pyrolysis oil • Gaseous fuel (Syngas)
3	Bio-chemical processing	• Anaerobic digestion • Fermentation • Landfilling and bioreactors	• Gaseous fuel (Biogas) • Biodiesel • Biofuel

Table 2.
Energy recovery of biomass [51–53].

the fossil fuel to minimize environmental impacts and air pollution, in particular. However, attention must be given to use the suitable energy conversion technique for better outcomes [52, 53].

4.3 Wind energy as clean energy resource

A very high energy potential is possessed by winds blowing at high velocities. This is because of the natural convective loops caused by the temperature difference in different regions. Due to this factor, this is considered as a by-product of solar energy. The high speed and impact of wind is converted into mechanical energy and then to electricity by using wind turbines. Various designs of wind turbines and mills are used for this purpose, depending on the available potential of winds in a particular region. Primarily, wind turbines are classified into 1—Horizontal Axis Wind Turbines and 2—Vertical Axis Wind Turbines. In the first category, the wind turbine rotates in the horizontal plane and it is governed by lift force of the winds. On the other hand, Vertical Axis Wind Turbines rotate in the vertical plane by drag force of winds. There are certain drawbacks and benefits for both the categories. However, it is important to check the feasibility of both designs for the desired location. The speed of the wind can change a lot over time, which can make it hard to rely on wind energy alone to meet energy needs. Therefore, a stable and reliable energy supply requires energy storage systems or backup power sources [54, 55]. The feasibility of optimum design of wind turbine can be achieved by using different simulation tools such as CFD, which can provide a true insight of thrust and impact of winds on solid body of the turbines. Based on such simulation results, the turbine height can also be adjusted for maximum output. Many countries have set high goals for generating renewable energy. Wind energy is expected to continue to play a big role in reducing air pollution and slowing climate change in the future. Modern energy generation systems particularly at seashore are equipped with such technologies for harvesting of maximum energy from winds. As per an estimate, this technology is expected to meet around 18% of the world's electricity demand by 2025 [56, 57].

4.4 Geothermal energy as clean energy resource

Geothermal energy refers to the energy extracted from earth core. The major benefit of this technique is the continuous and persistent potential of energy as compared to the other renewable energy resources. The high temperature in the earth core is extracted with the help of drilled pipelines filled with thermal fluid, the most commonly water. Sometimes, earth is taken as heat sink to increase or decrease ambient temperature. This is based on the concept that earth's temperature beneath the earth's surface remains constant throughout the year. Depending upon the local climatic condition, if a fluid is allowed to be injected in the earth surface with the help of heat conductor material (i.e., copper pipe), this will help to maintain the temperature as per earth's temperature. However, optimum depth, length, and orientation of pipe and the ambient weather conditions are the important factors influencing this technique. Buildings equipped with such technique will reduce the energy consumption for both cold and hot climatic conditions [58–61].

4.5 Hydro energy as clean energy resource

Hydro energy also sometimes termed as hydropower refers to the energy extraction from the kinetic energy possessed in the water. This is most suitable for hilly areas or natural water retaining locations where water has sufficient potential energy. In other words, a large volume of water is captured at a high altitude from where it is allowed to pass through a narrow path resulting in a very high speed of water stream. The hydro turbine is placed in this path, which converts the kinetic energy of water to electrical energy [62]. It is a source of energy that has been used for centuries, but in recent decades, it has been increasingly popular to reduce the emissions of greenhouse gases and ultimately mitigate air pollution. According to the International Hydropower Association, adoption of this technology, that is, hydropower contributes to prevent around 4 billion tons of CO_2 emissions every year across the globe [63, 64]. This is correlated with removing more than 1 billion cars from the road. There is a wide range of sizes and configurations available for hydropower plants, and they can be constructed on rivers, streams, or other bodies of water. Small-scale hydropower plants are only capable of producing a few kilowatts of electricity, whereas large-scale hydropower plants can produce thousands of megawatts [65].

5. Conclusions

Energy sector needs particular attention to mitigate environmental pollution, particularly for the case of air segment. Environmental degradation, because of consumption of fossil fuels and other primary energy resources in energy sector, has become critical challenge for the today's world. The cost of affordable energy is also growing drastically, which ultimately affects the overall per capita income. The air quality deterioration has also become very complex due to introduction of various new industrial processes and operations. Use of renewable and clean energy resources provides sustainable solution to both problems, i.e., fulfilling the energy requirements as well as mitigation of environmental pollution. Having natural cycle and occurring in the form of natural flux, such techniques have either zero or minimal impacts on the environment.

The future trends for the mitigation of air pollution in line with energy needs are summarized as below:

- Promotion of friendly policies and subsidies for the successful adoption of clean energy resources.
- Use of electric cars and vehicles powered by clean energy resources in the transport sector.
- Use of catalytic converters and filtration mechanism in the existing vehicles to avoid toxic emissions in the air.
- Utilization of modern IOT and AI tools for monitoring and predicting air quality so that real-time data can be collected, which could help researchers and policy-makers to make appropriate solutions for sustainable environment.
- Implementation of environmental laws and effluent standards particularly in energy sector.
- Awareness among the community about short-term as well as long-term effects of air pollutants to make and adopt community scale preventive measures.
- Precision farming and proper management of agricultural residues to avoid open burning of residues, wastes, and organic matters.

Conflict of interest

There is no conflict of interest.

Notes/thanks/other declarations

The authors gratefully pay thanks to the IntechOpen for providing an opportunity to publish this chapter free of cost.

Author details

Muhammad Usman Farid[1*], Atta Ullah[1], Abdul Ghafoor[2], Shahbaz Nasir Khan[1], Mazhar Iqbal[3], Furqan Muhayodin[2], Abdul Shabbir[4], Chaudhry Arslan[1] and Abdul Nasir[1]

1 Department of Structures and Environmental Engineering, University of Agriculture, Faisalabad, Pakistan

2 Department of Farm Machinery and Power, University of Agriculture, Faisalabad, Pakistan

3 University of Agriculture, Faisalabad Sub-Campus Burewala, Burewala, Pakistan

4 Department of Irrigation and Drainage, University of Agriculture, Faisalabad, Pakistan

*Address all correspondence to: engr.usman@uaf.edu.pk

IntechOpen

© 2023 The Author(s). Licensee IntechOpen. This chapter is distributed under the terms of the Creative Commons Attribution License (http://creativecommons.org/licenses/by/3.0), which permits unrestricted use, distribution, and reproduction in any medium, provided the original work is properly cited.

References

[1] Brunekreef B, Holgate ST. Air pollution and health. The Lancet. 2002;**360**:1233-1242

[2] Watson JT, Gayer M, Connolly MA. Epidemics after natural disasters. Emerging Infectious Diseases. 2007;**13**:1

[3] Tai AP, Martin MV, Heald CL. Threat to future global food security from climate change and ozone air pollution. Nature Climate Change. 2014;**4**:817-821

[4] D'amato G, Cecchi L, D'amato M, Liccardi G. Urban air pollution and climate change as environmental risk factors of respiratory allergy: An update. Journal of Investigational Allergology & Clinical Immunology. 2010;**20**:95-102

[5] Ukaogo PO, Ewuzie U, Onwuka CV. Environmental pollution: Causes, effects, and the remedies. In: Microorganisms for Sustainable Environment and Health. United States: Elsevier; 2020. ISBN: 978-0-12-819001-2

[6] He X, Liu Y. The public environmental awareness and the air pollution effect in Chinese stock market. Journal of Cleaner Production. 2018;**185**:446-454

[7] Builtjes P, Paine R. The problem–air pollution. In: AIR QUALITY MODELING-Theories, Methodologies, Computational Techniques and Available Databases and Software. USA: Air & Waste Management Association; 2003. p. 1

[8] Daly A, Zannetti P. An introduction to air pollution–definitions, classifications, and history. In: Zannetti P, Al-Ajmi D, Al-Rashied S, editors. Ambient Air Pollution. The Arab School for Science and Technology (ASST) and The EnviroComp Institute; 2007. pp. 1-14. Available from: http://www.arabschool.org.sy and http://www.envirocomp.org/

[9] Tang L, Xue X, Qu J, Mi Z, Bo X, Chang X, et al. Air pollution emissions from Chinese power plants based on the continuous emission monitoring systems network. Scientific Data. 2020;**7**:325

[10] Cole MA, Elliott RJ, Shimamoto K. Industrial characteristics, environmental regulations and air pollution: An analysis of the UK manufacturing sector. Journal of Environmental Economics and Management. 2005;**50**:121-143

[11] Phairuang W, Hata M, Furuuchi M. Influence of agricultural activities, forest fires and agro-industries on air quality in Thailand. Journal of Environmental Sciences. 2017;**52**:85-97

[12] Colvile R, Hutchinson EJ, Mindell J, Warren R. The transport sector as a source of air pollution. Atmospheric Environment. 2001;**35**:1537-1565

[13] Ghose MK, Majee S. Sources of air pollution due to coal mining and their impacts in Jharia coalfield. Environment International. 2000;**26**:81-85

[14] Strizhenok AV, Ivanov AV. Monitoring of air pollution in the area affected by the storage of primary oil refining waste. Journal of Ecological Engineering. 2021;**22**:60-67

[15] Bihałowicz JS, Rogula-Kozłowska W, Krasuski A. Contribution of landfill fires to air pollution–an assessment methodology. Waste Management. 2021;**125**:182-191

[16] Ying Y, Ma Y, Li X, Lin X. Emission and migration of PCDD/fs and major air pollutants from co-processing of sewage

sludge in brick kiln. Chemosphere. 2021;**265**:129120

[17] Rao ND, Kiesewetter G, Min J, Pachauri S, Wagner F. Household contributions to and impacts from air pollution in India. Nature Sustainability. 2021;**4**:859-867

[18] Zhang JJ, Samet JM. Chinese haze versus Western smog: Lessons learned. Journal of Thoracic Disease. 2015;**7**:3

[19] Arif F. SMOG: Causes, effects and preventions. Annals of King Edward Medical University; 2016;**22**:338-339. DOI: 10.21649/akemu.v22i4.1456

[20] Liu T, Duan F, Ma Y, Ma T, Zhang Q, Xu Y, et al. Classification and sources of extremely severe sandstorms mixed with haze pollution in Beijing. Environmental Pollution. 2023;**322**:121154

[21] Sastry N. Forest fires, air pollution, and mortality in Southeast Asia. Demography. 2002;**39**:1-23

[22] Halliday TJ, Lynham J, De Paula Á. Vog: Using volcanic eruptions to estimate the health costs of particulates. The Economic Journal. 2019;**129**:1782-1816

[23] Baluch MA, Hashmi HN. Investigating the impact of anthropogenic and natural sources of pollution on quality of water in upper Indus Basin (UIB) by using multivariate statistical analysis. Journal of Chemistry. 2019;**2019**:1-13

[24] Longhurst J, Barnes J, Chatterton T, Hayes ET, Williams W. Progress with air quality management in the 60 years since the UK clean air act, 1956. Lessons, failures, challenges and opportunities. International Journal of Sustainable Development and Planning. 2016;**11**:491-499

[25] Wilson WE, Suh HH. Fine particles and coarse particles: Concentration relationships relevant to epidemiologic studies. Journal of the Air & Waste Management Association. 1997;**47**:1238-1249

[26] Environmental Protection Agency, E. Particulate Matter (PM) Basics. United States: United States Environmental Protection Agency; 2022. Available from: https://www.epa.gov/pm-pollution/particulate-matter-pm-basics

[27] Zhang L, Yang Y, Li Y, Qian ZM, Xiao W, Wang X, et al. Short-term and long-term effects of PM2. 5 on acute nasopharyngitis in 10 communities of Guangdong, China. Science of the Total Environment. 2019;**688**:136-142

[28] Kelishadi R, Poursafa P. Air pollution and non-respiratory health hazards for children. Archives of Medical Science. 2010;**6**:483-495

[29] Kappos AD, Bruckmann P, Eikmann T, Englert N, Heinrich U, Höppe P, et al. Health effects of particles in ambient air. International Journal of Hygiene and Environmental Health. 2004;**207**:399-407

[30] Boschi N. Defining an educational framework for indoor air sciences education. In: Education and Training in Indoor Air Sciences. Dordrecht: Springer; 1999. pp. 3-6

[31] Bezirtzoglou E, Alexopoulos A. Ozone history and ecosystems: A goliath from impacts to advance industrial Benefits and interests, to environmental and therapeutical strategies. Ozone Depletion, Chemistry and Impacts (New York: Nova Science Publishers, Inc.). 2009:135-145. ISBN 978-1-61470-573-4

[32] Villányi V, Turk B, Batic F, Csintalan Z. Ozone pollution and its bioindication. In: Air pollution. Vol. 153. London, UK: IntechOpen;

2010. DOI: 10.5772/10047. ISBN: 978-953-307-143-5

[33] Lorenzini G, Saitanis C. Ozone: a novel plant "pathogen". In: Abiotic Stresses in Plants. Dordrecht: Springer; 2003. pp. 205-229

[34] Mccarthy JT, Pelle E, Dong K, Brahmbhatt K, Yarosh D, Pernodet N. Effects of ozone in normal human epidermal keratinocytes. Experimental Dermatology. 2013;**22**:360-361

[35] Gryparis A, Forsberg B, Katsouyanni K, Analitis A, Touloumi G, Schwartz J, et al. Acute effects of ozone on mortality from the "air pollution and health: a European approach" project. American Journal of Respiratory and Critical Care Medicine. 2004;**170**:1080-1087

[36] Woodward A, Smith KR, Campbell-Lendrum D, Chadee DD, Honda Y, Liu Q, et al. Climate change and health: On the latest IPCC report. The Lancet. 2014;**383**:1185-1189

[37] Environmental Protection Agency, E. Summary of the Clean Air Act [Online]. United States: United States Environmental Protection Agency; 2023. Available from: https://www.epa.gov/laws-regulations/summary-clean-air-act

[38] Hesselmann G, Rivas M. What are the main NOx formation processes in combustion plant. In: IFRF Online Combustion Handbook. File 66. Sheffield: International Flame Research Foundation (IFRF); 2001. Available from: https://ifrf.net/research/. ISSN 1607-9116

[39] Pielke RA Jr. What is climate change? Energy & Environment. 2004;**15**:515-520

[40] Jacob DJ, Winner DA. Effect of climate change on air quality. Atmospheric Environment. 2009;**43**:51-63

[41] Stocker TF, Qin D, Plattner G-K, Tignor MM, Allen SK, Boschung J, et al. Climate change 2013: The physical science basis. In: Contribution of Working Group I to the Fifth Assessment Report of IPCC the Intergovernmental Panel on Climate Change. United Nations Switzerland: Intergovernmental Panel on Climate Change; 2014. ISBN 978-92-9169-138-8

[42] Melamed ML, Schmale J, Von Schneidemesser E. Sustainable policy—Key considerations for air quality and climate change. Current Opinion in Environment Sustainability. 2016;**23**:85-91

[43] Thomas ER. Advanced nuclear energy: The safest and most renewable clean energy. Current Opinion in Chemical Engineering. 2023;**39**:100878. DOI: 10.1016/j.coche.2022.100878

[44] Goldemberg J. The promise of clean energy. Energy Policy. 2006;**34**:2185-2190

[45] Brown MA, Levine MD, Short W, Koomey JG. Scenarios for a clean energy future. Energy Policy. 2001;**29**:1179-1196

[46] Ge Y, Zhi Q. Literature review: The green economy, clean energy policy and employment. Energy Procedia. 2016;**88**:257-264

[47] Nwaigwe K, Mutabilwa P, Dintwa E. An overview of solar power (PV systems) integration into electricity grids. Materials Science for Energy Technologies. 2019;**2**:629-633

[48] Towoju OA, Oladele OA. Electricity generation from hydro, wind, solar and the environment. Engineering and Technology Journal. 2021;**39**:1392-1398

[49] Rajput SK. Solar Energy Fundamentals, Economic and Energy

Analysis. Rajnagar, Ghaziabad: Northern India Textile Research Association; 2017

[50] Alimujiang A, Jiang P. Synergy and co-benefits of reducing CO2 and air pollutant emissions by promoting electric vehicles—A case of Shanghai. Energy for Sustainable Development. 2020;**55**:181-189

[51] Kalak T. Potential use of industrial biomass waste as a sustainable energy source in the future. Energies. 2023;**16**:1783

[52] McKendry P. Energy production from biomass (part 2): Conversion technologies. Bioresource Technology. 2002;**83**:47-54

[53] Singh A, Basak P. Economic and environmental evolution of rice straw processing technologies for energy generation: A case study of Punjab, India. Journal of Cleaner Production. 2019;**212**:343-352

[54] Blanco MI. The economics of wind energy. Renewable and Sustainable Energy Reviews. 2009;**13**:1372-1382

[55] Doerffer P, Doerffer K, Ochrymiuk T, Telega J. Variable size twin-rotor wind turbine. Energies. 2019;**2019**(12):2543. DOI: 10.3390/en12132543

[56] Citaristi I. International Energy Agency—IEA. The Europa Directory of International Organizations 2022. Routledge; 2022

[57] Bilendo F, Meyer A, Badihi H, Lu N, Cambron P, Jiang B. Application and modelling techniques of wind turbine power curve for wind farms- a review. Energies. 2023;**16**:180

[58] Olabi AG, Mahmoud M, Soudan B, Wilberforce T, Ramadan M. Geothermal based hybrid energy systems, toward eco-friendly energy approaches. Renewable Energy. 2020;**147**:2003-2012

[59] Paulillo A, Cotton L, Law R, Striolo A, Lettieri P. Geothermal energy in the UK: The life-cycle environmental impacts of electricity production from the united downs deep geothermal power project. Journal of Cleaner Production. 2020;**249**:119410

[60] Adalı Z, Dinçer H, Eti S, Mikhaylov A, Yüksel S. Identifying new perspectives on geothermal energy investments. In: Multidimensional Strategic Outlook on Global Competitive Energy Economics and Finance. Emerald Publishing Limited; 2022

[61] Assad MEH, Zubayda SRM, Khuwaileh B, Hmida A, Alshabi M. Geothermal energy as power producer. Energy Harvesting and Storage: Materials, Devices, and Applications XI. 2021;**11722**:74-82

[62] Gernaat DE, Bogaart PW, Vuuren DPV, Biemans H, Niessink R. High-resolution assessment of global technical and economic hydropower potential. Nature Energy. 2017;**2**:821-828

[63] Gyanwali K, Komiyama R, Fujii Y. Representing hydropower in the dynamic power sector model and assessing clean energy deployment in the power generation mix of Nepal. Energy. 2020;**202**:117795

[64] Mitrovic D, Chacón MC, García AM, Morillo JG, Diaz JAR, Ramos HM, et al. Multi-country scale assessment of available energy recovery potential using micro-hydropower in drinking, pressurised irrigation and wastewater networks, covering part of the eu. Watermark. 2021;**13**:899

[65] Gokhale P, Date A, Akbarzadeh A, Bismantolo P, Suryono AF, Mainil AK, et al. A review on micro hydropower in Indonesia. Energy Procedia. 2017;**110**:316-321

Chapter 3

Phytoremediation toward Air Pollutants: Latest Status and Current Developments

Mahinder Partap, Diksha Sharma, Deekshith HN, Anjali Chandel, Meenakshi Thakur, Vipasha Verma and Bhavya Bhargava

Abstract

In recent years, air pollution has become one of the major environmental concerns that threaten health of the living organisms and its surroundings. Increasing urbanization, industrialization, and other anthropogenic activities impaired the air quality of indoor and outdoor environment. However, global organizations are focusing on ecological and biological means of solutions to reduce or eliminate dangerous contaminants from ecosystems in a sustainable manner. In this fact, plants are capable of improving or cleansing air quality and reduce the concentration of harmful pollutants from the environment through various remediation processes. Plants interact with air pollutants and fix them through various biological mechanisms in both associated and non-associated forms of microbes. In association forms, the mutualistic interaction of plant and microbes leads to higher growth efficiency of plants and results in enhanced pollutant degradation in rhizosphere as well as phyllosphere. In this background, the book chapter provides a comprehensive discussion of the existing literature and recent advances in phytoremediation process for the mitigation of harmful air pollutants. The role of indoor plants and aids for the enhancement of phytoremediation process towards air pollutants are also discussed.

Keywords: air pollution, phytoremediation, microbes, indoor gardens, biotechnological advancement

1. Introduction

Air pollution is one of the major threat to ecosystem services and imposes negative impacts on all living organisms [1]. The composition and type of air pollutants in a particular region majorly depends upon its sources, emission rate, and climate conditions. The pollutants that contribute to air pollution are as follows: carbon monoxide (CO), lead (Pb), nitrogen oxides (NO_x), ground-level ozone (O_3), sulfur oxides (SO_x), particulate matter (PM), volatile organic compounds (VOCs), poly-aromatic hydrocarbons (PAHs), and polychlorinated biphenyls (PCBs) [2–4]. Ground-level ozone is a colorless secondary pollutant and is produced when NO_x and VOC_s react in sunlight and stagnant air [5]. On the other hand, particulate matter (PM) is comprised of carbonaceous particles with accompanying adsorbed organic substances and reactive metals [6]. Gaseous pollutants such as SO_x and NO_x also aid particle formation through complex atmospheric photochemical reactions involving ammonia released from agricultural fields [5]. Road traffic is a significant source of NO_2, particularly from diesel automobiles that contribute to the global air pollutants emission in large cities. However, the main sources of SO_2 are industrial emissions and maritime transportation.

Global organizations are concerned about the increasingly deteriorating air quality because it is believed to be one of the leading causes of approximately 3.1 million fatalities each year [7–10]. The average life expectancy of urban populations is decreasing as exposure to high levels of air pollutants over a longer period of time. Pollutants like fine particulate matter ($PM_{2.5}$) and ground-level ozone are unquestionably linked to an increase in mortality rate [11]. Particulate matter alone causes23% of total damage to human health despite representing only 6% of the total air pollutants [12]. Pollution-related illness has been estimated to cause as many as 9 million premature deaths during 2015, which was three times greater than the number of deaths from other illnesses such as malaria, AIDS, and tuberculosis altogether [13]. Alongside, the air quality of the indoor environment has become a global issue as people in urban areas spend more than 90 percent of their day in office spaces or residential areas [14]. Poor indoor microclimate adversely affects the health, happiness, and productivity of occupants [15]. Major indoor air pollutants such as PMs, VOCs, PAHs, PCBs, NH_3, SO_2, H_2O_2, HNO_3, HNCO, CO, and H_2S upon inhalation are linked to a range of health problems including asthma, heart disease, reproductive problems, neurological issues, irritate the eyes, and respiratory disorder [13].

Remediation of these pollutants for sustaining ecosystem and human health using either physical, chemical, or biological approaches is applicable but limited due to the cost, labor requirements, and safety hazards [16]. Here, phytoremediation can be effectively used as an alternative technique, and it is gaining popularity, acceptance, and implementation due to ease and an array of benefits. Phytoremediation is the use of plants and associated microorganisms to reduce or degrade the concentrations or toxicity of pollutants. Phyto-stabilization, rhizo-degradation, phyto-extraction, phyto-degradation, phyto-volatilization, and phyto-filtration are the fundamental mechanisms of the phytoremediation process. The efficiency of these processes can be improved by using synthetic and natural additives, suitable microbes and host, and genetic engineering/editing. Various modes of biological remediation include microbes-related remediation, enzyme-assisted remediation, vermi-remediation, phyto-remediation, and zoo-remediation exists in nature. However, microbial-assisted phytoremediation including plants and microbes is one the most effective, sustainable, and economical approach to reduce harmful pollutants from the environment [17, 18]. The bioremediation of air pollutants by the phyllosphere or microbes associated with the leaves, not just the microbes themselves, is known as phyllo-remediation. While rhizo-remediation is the process of degrading organic contaminants in the soil region around plant roots (the rhizosphere), typically as a consequence of enhanced catalytic activities of root associated microbes [19–21].

Various ornamental plant species have been used for the enhancement of air quality. Previously, *Chamaedorea elegans* and *Opuntia microdasys* plants have been shown to reduce the concentrations of formaldehyde and BTEX. Also, *Chlorophytum comosum* L. plants are able to accumulate indoor particulate matter pollutant (PM_{10}, $PM_{2.5}$, and $PM_{0.2}$). The rhizospheric microbes associated with the *Aloe vera, Tradescantia zebrina*, and *Vigna radiata* plant species have improved or enhanced the formaldehyde neutralization efficiency by 2–3 times [22]. The vertical gardens growing in soil-less system, the nutrient-enriched solution with activated carbon have reduced the level of VOCs in indoor environments [23]. Considering this, the book chapter discusses the existing literature and recent advances in the phytoremediation process and its potential against air pollutants.

2. Air pollutants: types, sources, and health issues

Gaseous pollutants and particulate matter are the two primary types of pollutants that are found in the atmosphere (**Figure 1**). There is a substantial difference between the levels of these pollutants found in outdoor and indoor environments which is primarily allocated to the sources from which they are derived.

2.1 Carbon dioxide (CO_2)

Carbon dioxide (CO_2) is a colorless and odorless gas, and it constitutes about 0.03% (300 parts per million) of the total atmospheric gases. It is heavier than other

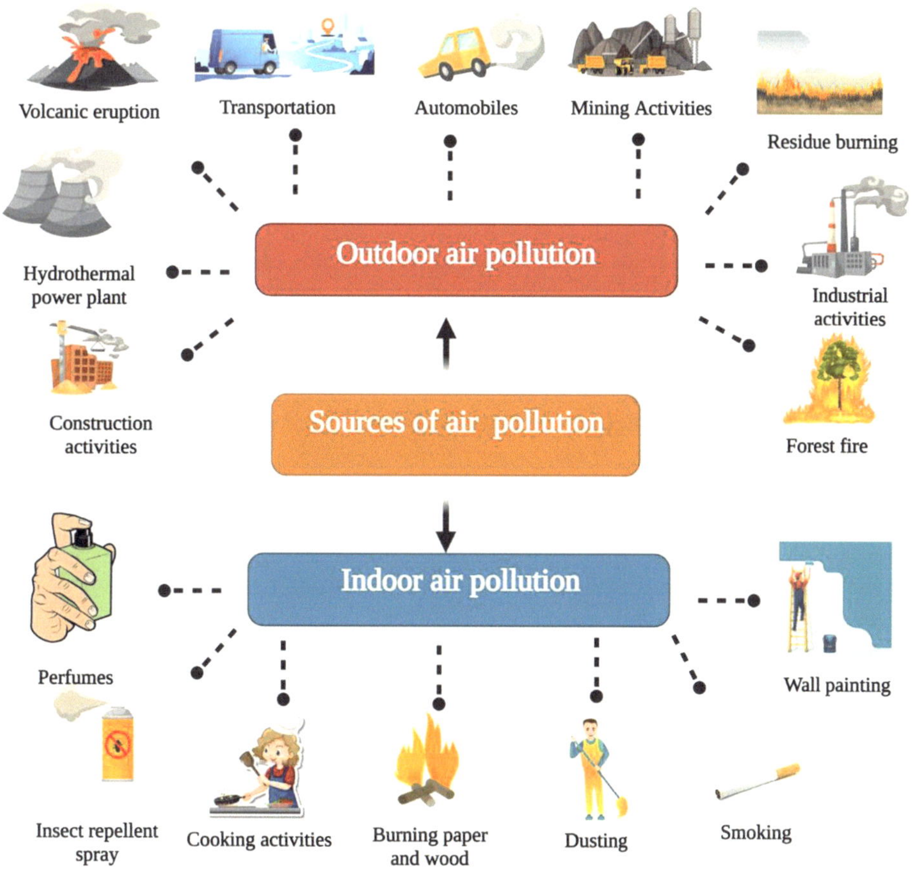

Figure 1.
Different sources of outdoor and indoor air pollutants.

non-combustible gas and can accumulate in the lower phase of the environment, resulting in an oxygen deficiency [24]. Carbon dioxide is majorly a by-product of biological respiration or fossil fuel combustion [25, 26]. Fossil fuel-fired power plants contribute about 33–40% of the total CO_2 emission globally, with coal-fired power plants being the key contributor [27]. Whereas, anthropogenic CO_2 emission is due to activities like forestry, deforestation, land clearing for agriculture [28]. Rising atmospheric CO_2 levels result in the greenhouse effect and speed up the global warming process [29]. The average ambient CO_2 concentration has been steadily rising and has reaches up to 410 ppm. A CO_2 level of 600 ppm is considered acceptable, but a CO_2 level above 1000 ppm is detrimental and leads to CO_2 toxicity-related side effects [30]. Excessive CO_2 concentration in the blood (hypercapnia) give rise to acidosis, which is characterized by a low blood pH (increased acidity). The respiratory, cardiovascular, and central nervous systems are all affected by the lower blood and tissue pH. Other commonly reported symptoms of CO_2 toxicity are headaches, lethargy, moodiness, mental slowness, emotional irritation, and sleep disruption [31].

2.2 Carbon monoxide (CO)

Carbon monoxide is colorless, combustible, and extremely deadly gas [32]. It is emitted from both natural and man-made sources. It is produced when carbonaceous materials are burned in an incomplete manner [33]. The two most common sources of emission of carbon monoxide in ambient air are smoke from fires and exhaust fumes from automobile engines (in the absence of a catalytic converter) [34]. Additionally, the combustion of charcoal and wood can also release carbon monoxide. The minimum exposure limit of carbon monoxide is around 100 mg/m^3 for 15 minutes, 60 mg/m^3 for 30 minutes, 30 mg/m^3 for 1 hour, and around 10 mg/m^3 for 8 hours [35]. In most cases, carbon monoxide poisoning causes from inhalation of gases coming out from common household areas such as garages, kitchens, basements, or workrooms and fuel burning. Carbon monoxide toxicity leads to dizziness, headache, weakness, nausea, vomiting, and loss of consciousness [36]. Further, the inability of a cell to use oxygen (e.g., effective oxygen deprivation) leads to chemical asphyxiation and hypoxia, which is the most significant harmful effect of carbon monoxide.

2.3 Volatile organic compounds (VOCs)

VOCs are categorized as harmful air pollutants because of their relation with carcinogenic, impairing blood production and weakening nervous system in humans [34]. Commonly found VOCs in atmosphere are aromatic hydrocarbons BTEX (benzene, toluene, ethyl benzene, and xylene) and halogenated hydrocarbon like chloroethylene and trichloroethylene [37]. The major sources of BTEX are vehicle/aircraft, processing of petroleum products, paints, thinner, ink, cosmetics, and pharmacy. Benzene/toluene ratio is used more often to know the source of emission. If the ratio is >0.5 shows that the source is other than transport and if the ratio is <0.5, vehicular emission is the major source. Among the six compounds of BTEX, toluene is the one that is most easily degraded due to the presence of a side chain that provides different attack routes for the microbial enzymes to act upon. Isoprene, a naturally occurring biological component, is one of the most significant contributors to emissions of VOC [38]. Anthropologically, VOCs are generated from both domestic and industrial processes including textile cleaning, fertilizers and pesticide application, septic system, traffic, fumigation, building materials, and pharmaceutical industries [39]. In indoor environments, VOCs are released by combustion, newly constructed or refurbished structures and building materials such as paints, carpets, solvents, various plastics, and wooden furniture [40]. The acceptable level of VOCs concentration in an indoor environment is ranged from 0 to 400 ppb [41, 42]. Short-term exposure of VOCs may induce like nausea, vomiting, and fatigue. However, long-term exposure may cause lung cancer, leukemia, and other forms of malignancy [43, 44].

2.4 Particulate matters (PMs)

Carbon-containing particles along with reactive metals and adsorbed organic compounds belong to the particulate matter components of air pollution [7]. It is a mixture of solid and liquid particles in the air that can be breathed in and may cause serious health problems. According to particle size, PM is further divided into three groups: coarse particles (PM_{10}, diameter less than 10 μm), fine particles ($PM_{2.5}$, diameter less than 2.5 μm), and ultra-fine particles ($PM_{0.1}$, diameter less than 0.1 μm) [7]. It can be originating either from natural or anthropogenic activities. Eruptions of volcanoes, dust and wind storms, forest fires, salt spray, rock debris, chemical reactions between gaseous emissions, and soil erosion are some examples of natural sources. PM is also produced by human activities such as burning fuel, making steel, processing petroleum, making cement, making glass, mining, smelting, power plant emissions, burning coal, and disposing of agricultural waste [45]. There is a clear relationship between PM concentrations and seasonal variations [46, 47]. According to the recent air quality guidelines 2021 by WHO, exposure to $PM_{2.5}$ and PM_{10} concentration up to 65 $\mu g/m^3$ for the 24-hour is safe. Black carbon is a carbonaceous component released because of incomplete combustion of fossil fuels (particularly diesel, wood, and coal) (PM 2.5) [10, 48]. Higher exposure to black carbon is a major health issue that can induce heart attacks and strokes. The regulations governing air pollution focus on $PM_{2.5}$ as the primary concern. PM_{10} mostly affects the upper respiratory system, whereas ultra-fine particles detrimental to the lower respiratory tract, lungs, and alveoli.

2.5 Polycyclic aromatic hydrocarbons (PAHs)

These are the vast categories of chemical compounds that include two or more bound benzene rings as diverse arrangements [49]. PAHs are a type of pollutant that can be found almost everywhere in the environment including soil, water, and air. PAHs are the by-product of the incomplete burning of organic substances such as wood, coal, petrol, and oil [44]. Forest fires, garbage incineration, volcanic eruptions, and hydrothermal processes are natural sources of PAHs. Whereas, combustion of timber, waste, and fossil fuels are some examples of anthropogenic activities which are responsible for the emission of PAHs [45, 48]. PAHs have cancer causing and mutation inducing properties in living organisms [47, 50]. The most common ways for people to be exposed to PAHs are through smoking cigarettes or cigars and breathing smoke from open fires or other sources of combustion [46]. Health issues such as skin-related diseases, lungs and gastrointestinal malignancies, and damages in liver and loss of immunity have been resulting in long-term exposure. The National Institute for Occupational Safety and Health (NIOSH) recommends that exposure to PAHs in the workplace should be limited at or below the minimum reliable detectable concentration of 1 $\mu g/m^3$.

2.6 Ozone (O_3)

Ground-level ozone is produced by a chemical interaction between oxides of nitrogen and originating from natural sources and/or as a result of human activities. Ozone damages the uppermost skin layers and tear ducts. Short-term exposure is resulted in malondialdehyde production in the epidermis as well as vitamin C and E depletion in mice model [1]. Due of ozone's limited solubility in water, it can enter deep into the lungs when inhaled. During the warm season, an increase in ozone concentration was related to an increase in the daily number of fatalities (0.33%), respiratory deaths (1.13%), and cardiovascular deaths (0.45%). During the winter, no such effect was noticed [51].

2.7 Nitrogen and sulfur oxides

Nitrogen oxides (NOx) are the gases that are released by natural sources, automobiles, and other fuel-burning actions [1]. NO_2 is an odorous, acidic, and extremely corrosive gas that have negative impact on human health and the environment. They are responsible for the yellowish-brown hue of the smog. It causes pulmonary disorders such as obstructive lung disease, asthma, chronic obstructive pulmonary disease, and in rare circumstances acute aggravation of COPD as well as fatalities [52]. High concentrations of NO_2 are also detrimental to vegetation causing damage to leaves, stunting growth and diminishing crop yields. The suggested NO_2 air quality criteria are 0.12 ppm for a 1-hour exposure duration and 0.03 ppm for a yearly exposure period. These regulations are intended to safeguard vulnerable individuals such as children and asthmatics [53, 54]. Sulfur oxides (SOx) composed of molecules of sulfur and oxygen is an odorless gas detectable by taste and smell at concentrations between 1000 and 3000 micrograms per cubic meter [55, 56]. The majority of SO_2 is produced by the combustion of sulfur-containing fuels and metal sulfide ores. Volcanoes are natural sources of SO_2 (35–65%) among others. After industrial boilers and nonferrous metal smelters thermal power plants that burn high-sulfur coal or heating oil are generally the largest producers of anthropogenic SO_2 emissions on a global scale. The accumulation of SO_2 and smoke which reached 1500 mg/m^3 resulted in increased number of fatalities [55, 57].

3. Absorption of air pollutants in plants

3.1 Stomata

The primary entry site for air pollutants in the plants is likely to be stomatal pores, either through exchange of gases or penetration of a liquid layers into the sub-stomatal cavity. By altering the aperture between the guard cells, plants regulates absorption of air pollutants with a diverse range of molecular masses. The movement of guard cells is regulated by the K^+ concentration in the cell sap. However, the mechanism of stomatal response to external environmental stimuli involves cellular sodium potassium pump and calcium homeostasis of guard cells as well as K^+ ion and its counter-ions, malate, chloride, and nitrate [49, 58]. Several environmental factors (temperature, light, and relative humidity) and cell internal factors (partial pressure of CO_2, sucrose concentration, turgidity of guard cells and the presence or absence growth inhibitors) also affect stomatal opening. Absorption of oxides of sulfur (SOx) and nitrogen (NOx) is regulated by light intensity and heavy metals with airborne particulates will directly enter through the stomatal pores, while lipophilic substance enters through cuticular wax [59].

3.2 Cuticle

When stomata are closed, the cuticle acts as the entry point for the pollutants. Lipophilic substances enter through cuticles of the plants, whereas hydrophilic compounds such as gaseous and liquid contaminants can be absorbed by cuticle to some extent. The morphology and composition of cuticles vary with species age, and location of the cuticle in plant as well as some climatic factors such as temperature and relative humidity. Organic compounds such as PAHs, PCBs, dioxins, and charged particulates may alter chemical constituent of cuticle that increase permeability of the cuticle. Once these pollutants enter through cuticle, they infiltrate slowly by diffusion process or get deposited on cell wall or the vacuoles [60].

4. Phytoremediation and its mechanism

In phytoremediation, plants absorb contaminants from the surrounding atmosphere and then degrade or detoxify them using a variety of mechanisms [61, 62]. It is currently unclear how specific air pollutants are captured by plants for subsequent degradation, metabolization, or sequestration. Additionally, phytoremediation has been studied for plant propensity to filter ambient air and exchange gases with their surroundings [63]. The large biologically active surface areas of plants have added advantage in capturing different air pollutants through absorption, transport, and deposition of organic pollutants in the rhizosphere and phyllosphere [64]. For the elimination of air pollutants, the photosynthetic systems of C3, C4, crassulacean acid metabolism (CAM), and facultative CAM plants have been studied under various circumstances [65]. C4 plants posses high intensity in exchange of gases as compared to C3 plants and may exchange more CO_2, especially during the day. CAM plants exchanges the gases during the night which makes them highly efficient in phytoremediation specially when they are placed in indoors [66]. Microorganisms that make up a phyllo-microbiome colonize on leaf surfaces and have potential to degrade a variety of organic contaminants [67]. Even soil microorganisms have ability to eliminate gaseous air pollutants when they are associated with plants [68]. Various mechanisms in phytoremediation are discussed below (**Figure 2**).

Phytoextraction refers to the process of taking up of pollutants from soil and transporting them to above ground plant parts for further degradation. Thus, in phytoextraction both phyllosphere and rhizosphere are involved [69]. Efficacy of phytoextraction process is dependent upon mobility and availability of heavy metals in the root zone [70–72]. Phytovolatilization is transport of contaminants into the phyllosphere of plants through the epidermis by diffusion across the cuticle and also through open stomata. These contaminants are further degraded into volatile components which are further released in the atmosphere through transpiration from the stomata [73]. Transpired volatile components either stay in the atmosphere as an air pollutant or they may break down by the action of hydroxyl radicals [74]. Phytovolatilization process is observed for number of contaminants both inorganic and organic forms. In phytovolatilization, pollutants are absorbed by plants through phyllosphere and transformed into volatile compounds. Eventually, these degraded volatile substances transpired into the atmosphere via stomata [75].

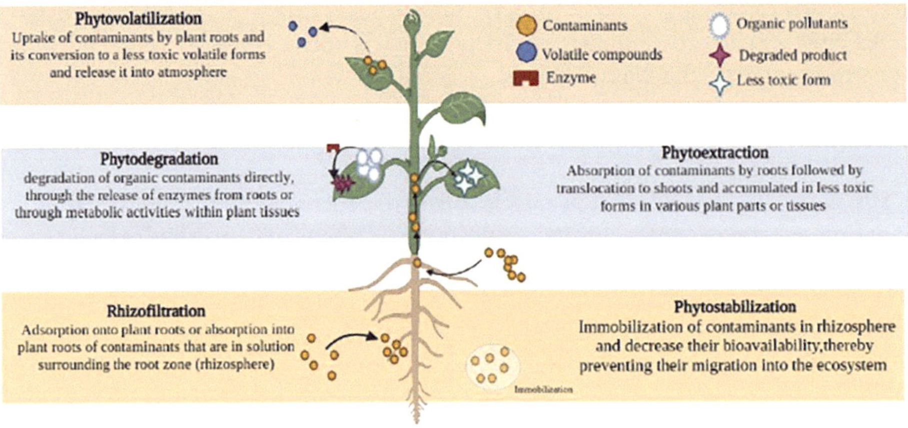

Figure 2.
Mechanisms of phytoremediation.

In phytodegradation, organic contaminants including polycyclic aromatic hydrocarbons (PAHs), total petroleum hydrocarbons (TPHs), polychlorinated biphenyls (PCBs), and inorganic contaminants like atmospheric nitrogen oxides and sulfur oxides are taken up and transformed by plants into simpler less toxic forms [9]. The obtained products of phytotransformation are incorporated into phyllosphere and rhizosphere. Specific plant enzymes such as nitroreductases, dehalogenases, laccases, and peroxidase play vital role in the phytotransformation process [73]. The immobilization of contaminants takes place in the rhizosphere during phytostabilization and is known as phytoimmobilization. Lignin, which is found in the cell wall of plant roots, is able to adsorb pollutants which then precipitates into insoluble compounds and stores them in the root zone [71]. Rhizodegradation refers to the biodegradation of contaminants in the soil by edaphic microbes enhanced by the inherent character of the rhizosphere itself. The roots of plants offer an additional surface area which allows for the transmission of oxygen and the growth of microorganisms. Plants emit biodegradable enzymes and metabolites into the rhizosphere, where microbes can use them to develop and break down contaminants [76]. Root exudates have many potential uses such as enhancing plant defense mechanism, boosting nutrient availability in the root zone, and even phytoextraction of heavy metals. The ability of bacteria to degrade pollutants is *Bacillus*, *Acinetobacter*, *Arthrobacter*, *Diaphorobacter*, *Enterobacter*, *Flavobacterium*, *Phanerochaete chrysosporium*, *Polysporus*, *Pseudomonas*, *Pseudoxanthomonas*, *Rhodococcus wratislaviensis*, *Sphingomonas*, and *Stenotrophomonas*.

5. Phytoremediation of major pollutants

5.1 Airborne particulates

By adsorbing particles leaf surface or fixing them in waxes, plants effectively remove substantial amounts of airborne particulates from the atmosphere, particularly in metropolitan areas [77]. Urban areas with trees can significantly reduce PM_{10} level [73]. In order to decrease the spread of air pollution from industrial areas, an

8 m wide green belt may be installed which could able to minimize dust fall over two to three times [78]. The trees such as *Ficus* spp., *Mangifera* spp., and *Azadirachta* spp. in urban roadsides can effectively control particulate matter emitted from vehicles. C_3 and C_4 plants intake several gaseous pollutants at larger quantity during daytime and CAM plants intake gases at night time through stomatal openings [79]. PCBs enter into plants through the cuticle and metabolized through cytochrome P-450 in some of the plant species such as pine and eucalyptus [80]. Generally, metals such as Cr, V, Ni, Pb, and Fe accumulated in plants from soil, but they are not translocated to aerial plant parts. Therefore, accumulation of such metals in aerial parts of the plants is majorly absorbed from atmospheric air [81]. High density organic compounds and metals can penetrate wax layer and enter into epidermis of the cell through the process of diffusion and sequestered in vacuoles or cell wall [82]. The ability of some bacteria to convert reactive oxygen species into less harmful forms through their anti-oxidant properties. This ability of bacteria in turn contributes to the bio-remediation of PM by plants as PM have shown to develop ROS which is harmful to plants [83–86].

5.2 Volatile organic compounds

Phytoremediation can effectively remove VOCs like formaldehyde, xylene, toluene, benzene, and ethylbenzene from the environment [87]. In order to protect from pathogens, animals and other environmental stresses plants used low molecular weight VOCs [84]. Generally, plant uptake VOCs via leaf stomata, while few plants uptake VOCs through cuticle [88], and their further translocation is done by phloem to designated plant organs [89]. In plants system, VOCs are degraded, stored or removed through various process, and get volatilized into atmosphere through diffusion from trunks, stems, roots, and leaves of plants [90, 91]. Phytoremediation efficiency of plants is determined by properties of VOCs, as lipophilic VOCs are absorbed through cuticle, whereas hydrophilic VOCs are absorbed through leaf stomata [13]. VOCs also get deposited in soil and plant rhizosphere due to leaf fall and runoff. Microbes present in plants aid metabolization of these organic compounds into less toxic forms such as carbon dioxide, water, and cellular biomass [13]. Spider plant (*Chlorophytum comosum* L.) detoxifies low concentrations of formaldehyde into amino acid, sugars lipids, and cell wall components. Soybean plant (*Glycine max* L.) converts formaldehyde into serine and phosphatidylcholine. Additionally, microbes can modify plant VOC emissions by activate an immunological response [92]. An experiment was conducted using the green wall system for degradation of VOCs and showed proteobacteria as the dominant species. *Nevskiaceae* and *Patuli bacteraceae* were VOC utilizing bacteria in the irrigation water of the green wall system. *Burkholderiales* were part of bio-wall root bacterial communities where VOC degradation was also reported [93]. Bacteria associated with the rhizosphere of the plants also play a crucial role in the degradation of these air pollutants, and one such bacteria *Rhodococcus erhythopolis* U23A isolated from the roots of *Arabidopsis thaliana* was able to break down polychlorinated biphenyls [94]. The flavanones were found to be the inducers of the polychlorinated biphenyls pathway. According to Barac et al. [94], the introduction of a plasmid expressing a toluene-degrading enzyme reduced phytotoxicity and toluene evapotranspiration by 50–70% through the leaves. Sandhu et al. [95] provided direct evidence that endophytic bacteria in the phyllosphere degrade volatile organic compounds. De Kempeneer et al. [96] proved that phyllosphere micro-biota undertakes toluene cleanup via toluene-degrading bacteria. Kempeneer et al. [96] demonstrated that phyllosphere micro-biota significantly degrades the toluene compound. As per the reports, formaldehyde has varied from 0.14 to 0.61 mg/m^3 in remodeled residences; however, benzene, toluene, and xylenes were found in 124, 258, and 189.7 g/m^3 concentrations, respectively [97]. Sriprapat et al. [98] highlighted that *Helianthus annuus* associated with EnL3 strain significantly removed benzene from the environment. The study also reported that a total of sixty-two proteins was up and down-regulated in leaves, while thirty-five proteins were up or down-regulated in roots of *H. annuus*. Additionally, toluene has been removed by the association of *F. verschaffeltii* and *H. carnosa* indoor foliage plants with rhizospheric bacteria [99, 100]. To summarize these, it is clear that the above ground and below ground plant-associated microorganisms play a crucial role in the mineralization of VOCs through their degradation potential.

5.3 Gaseous pollutants

Gaseous air pollutants include oxides of sulfur, carbon, and nitrogen as well as ozone [101]. Plants metabolize CO by oxidation into CO_2 or get reduced into amino acids. The tendency of plant canopy to act as NO_2 sinks and assimilation of NO_2 has also been demonstrated. Plants can assimilate ammonia (NH_4) from the air and soil [102]. Abatement of pollutants such as carbon dioxide (CO_2), sulfur dioxide (SO_2), nitrogen dioxide (NO_2), and ozone (O3) can be accomplished by the implementation of phytoremediation. Through the process of photosynthesis, plants remove carbon dioxide from the atmosphere and store it for any given period of time or convert it to humus [13]. This process of storing carbon dioxide in plants for an extended period of time is known as carbon sequestration [103]. The “reductive sulfur cycle” is the process by which sulfur dioxide is broken down after it is diffused through the stomata of plants. Plant’s root uptake by-products of the sulfur cycle including sulfur containing amino acids necessary for their growth. The adsorption of nitrogen dioxide through stomata, leaf, and root surfaces is the first step in the nitrogen toxicity abatement process. Nitrogen dioxide has the potential to be used as an alternative fertilizer and to supply essential nutrients to plants but, prolonged exposure to high concentrations of nitrogen within the plants could be toxic [13]. Ozone reacts with the waxes of cuticle, ions, salts, and biogenic and anthropogenic volatile organic compounds. Stomata are the means through which plants take in ozone into their systems. In another study, plant and microbially-assisted bio-remediation systems accrued the air pollutant by 0.24 to 9.53 folds than individual plant systems [104]. Moreover, ozone was efficaciously reduced or removed in an uninterrupted system of *Z. zamiifolia* combined with *B. cereus* ERBP. This endophytic bacterium has the potential to protect the plant against ozone stress [105] and has been developed an effective microbially-assisted bio-remediation strategy against formaldehyde pollutants by adhering rhizosphere microbes with *T. zebrina*, *A. vera*, and *V. radiata*. As per the report [49], the phenolic pollutant has been eliminated by vetiver grass by using *A. xylosoxidans*. The association of F3B has the potential to endure plants against toluene stress and enhanced the chlorophyll content of leaves. As per the report [106], chloromethane (volatile halo-carbon pollutant) was removed/degraded by *Hyphomicrobium* sp. (isolated from the leaf of *A. thaliana*). NO_2 may also be accumulated in plants in the form of nitrate and nitrite. Later, it can be reduced by nitrate and nitrite reductase enzyme for the generation of ammonia, which is further assimilated to glutamate through the ammonium assimilation pathway (GS- GOGAT) [107, 108]. As per the recent report by Lee et al. [16], the absorbed pollutants are stored in the inter-cellular spaces of leaves. Further, it can react with the inner-leaf membranes or water film and after that it can be degraded or excreted into the environment [16].

5.4 Removal of polyaromatic hydrocarbons (PAHs)

Various bacterial groups like *Haemophilus* spp., *Paenibacillus* spp., *Pseudomonas* spp., *Mycobacterium* spp., *and Rhodococcus* spp. are reported to degrade and utilize PAHs as a sole source of energy and carbon [109]. Yutthammo et al. [110] reported that about 1–10% of phyllosphere bacteria have potential to degrade PAHs (acenaphthylene, acenaphthene, phenanthrene, and fluorine) from the environment. However, the removal of PAH was achieved by the mutualistic association between microbes and *Epipremnum aureum* plant [111]. A lot of changes take place in the soil around the roots which leads to more aeration that supports the growth of autochthonous PAH degrading microbial communities, which allows the efficient mineralization of the PAH present in the rhizosphere soil [112]. Thus, even deeper layers support the process of aerobic degradation [113]. Myco-augmentation, phytoremediation and natural attenuation can be used individually for the bioremediation process. However, using these techniques in combination can increase bioremediation efficiency to several folds. Thus, the synergism between the microbes and plants can be exploited for the bioremediation not only for PAH but also other compounds as it is more effective than simple phytoremediation [114]. There have been few studies that has used the combination of bacteria and plants for the bioremediation of PAH [49], but rarely there has been a combination of plant and white-rot fungi used for the process of PAH bioremediation. The maize plant was associated with *Crucibulumleave* (fungi) in a comparative study where phytoremediation process was enhanced. This combination was highly effective in PAH degradation to 5–6 folds compared to phytoremediation alone. This could be possible due to the increased surface area of fungal hyphae which could assist bacterial transport through the soil and alteration of root exudates possibly increasing the bioavailability of the compounds and increasing the degradation of the PAH [115]. For the removal of hydrocarbons from soils, the most researched plants are prairie grasses because of their vast fibrous root systems [116].

5.5 Indoor air pollution

Green houseplants can act as a biological filter to purify the indoor air [76]. Plants considerably deplete CO_2, VOCs, PMs, organic carbon, nitrate, sulphate, ammonia, and carbonate levels in indoor environments [117]. It was reported that *Dracaena deremensis*, *Dracaena marginata*, and *Spathiphyllum* spp. efficiently remove benzene, toluene, ethylbenzene, and xylenes in indoor environment [118]. Eight ornamental indoor plants namely *Chlorophytum comosum*, *Clitoria ternatea*, *Dracaena sanderiana*, *Euphorbia milli*, *Epipremnum aureum*, *Hedera helix*, *Syngonium podophyllum and Sansevieria trifasciata* were studied for the removal efficiency of benzene in indoor air pollutants. It was found that *C. Comosum* was most efficient in removing benzene from air and water pollutants. Green walls are recent innovation, formed of a pre-vegetated frame that are attached to a wall or other interior structure [87, 117]. An updated version called active living wall integrates the building heating, cooling, and ventilation systems [119]. A green wall system regulates temperature and also filters the air inside buildings [120]. The plants remove CO and CO_2 and assist in removing particulate matter from air [121, 122]. Using urban indoor vegetation is one strategies for accomplishing this since it can be drastically reducing air pollution. Green walls are either partially or entirely covered with greenery, incorporating a growing medium. It is well-known that the incorporation of green walls and other forms of living infrastructure into an indoor environment has the potential to contribute to an improvement in air quality, through the reduction in amount of volatile organic compounds (VOC), inorganic gaseous compounds, and carbon dioxide (CO_2), as well as the retention of particulate matter (PM) [123]. Study revealed that potted plants are capable of removing significant levels of gaseous VOCs in sealed chambers, lowering VOCs by 10–90% in 24 hours [124]. It was demonstrated that the removal of organic contaminants is accomplished more effectively in areas of the root soil that are in contact with air [125]. The *Chamaedorea elegans* and *Opuntia microdasys* plants have significantly reduced the concentrations of formaldehyde and BTEX in controlled environmental chambers, respectively. In another study, the *Chlorophytum comosum* L. plants have significantly accumulated indoor PM10, PM2.5, and PM0.2. The researchers also observed that the plant wax is helpful in the accumulation or attachment of PM on their leaf's surfaces [126]. As per the literature, associated microorganisms can significantly enhance the remediation properties of plants. In one experiment, the rhizosphere microbes associated with the *Aloe vera*, *Tradescantia zebrine*, and *Vigna radiata* plant species have improved or enhanced the indoor formaldehyde neutralization efficiency by 2–3 times. In addition, the combined system of the *Ophiopogon japonicus* plant associated with *Staphylococcus epidermidis* and *Pseudomonas spp.,* bacterium has been degrading the phenol pollutants with 1000 g/L per day capacity [118]. In commercial buildings and urban areas, vertical gardens are gaining more attention because vertical alignment can offer a space-efficient method of exposing more plant biomass to contaminated air [127]. In these gardens, the significance of plant selection on the green walls demonstrated the varying capacity of pollutant degradation. As per the report, the especially fern species have high removal efficiency toward the particulate matter of size range $PM_{0.3–0.5}$; 45.78% and $PM_{5–10}$; 92.46%. Whereas, the plant species with fibrous roots have greater degradation efficacy toward air pollutants compared to plants with tap root systems. Moreover, for vertical gardens and green walls, the growing medium, vermicompost, perlite, coco-peat, and so forth significantly influence the plant microbe's associations and pollutant degradation mechanism. In *Apteniacordifoli*, the addition of granular activated carbon into coconut husk-based substrates could increase the VOC removal ability of the green walls [128]. Mikkonen et al. [129] investigated the filtration efficiencies for seven volatile organic chemicals as well as the microbial dynamics in simulated green wall systems. The results highlighted that Golden pothos plants had a minor effect on VOCs filtration and bacterial diversity. In another report by Gonzalez-Martin [92], indoor green walls contributed significantly to the ambient fungal load, but concentrations remained well below WHO safety standards. In this sense, the botanical air filtration approach is a promising way for reducing indoor air pollution, but a greater knowledge of the underlying mechanics is still required.

6. Biotechnological advances

Generally, biotechnological tools provide researchers the ability to change the gene expression regulation at certain specified sites which speeds up the discovery of new information regarding the functional genomics of plants [128]. In *Noccaeac aerulescens*, *Arabidopsis spp.* (hyperaccumulator of Cd and Zn), *Hirschfeld spp.* (tolerant to Pb toxicity), *Pterisvittata*, and *Brassicaspp* have genomes sequenced for model phytoremediators [129]. These phytoremediators have been found to be effective at removing heavy metals from the environment. Similarly expressing the *MerC* gene in *Arabidopsis* and Tobacco plants led to an increase in the accumulation of the Hg metal, but it also rendered the plant hypersensitive to the effects of mercury (Hg). Several additional types of organic contaminants such as polycyclic aromatic hydrocarbons and polychlorinated biphenyls are also include in this method [92]. Editing the genes in rice and *Arabidopsis* for the production of naphthalene dioxygenase resulted in the recent development of genes that express tolerance against naphthalene and phenanthrene [130]. Some of the genetically modified plants for phytoremediation of organic pollutants include tobacco (gene *hCYP2E1*) for benzene, *Atropa belladonna* (gene *rCYP2E1*), and Poplar (gene *rCYP2E1*) for TCE [131]. An experiment was carried out on the *tou* cluster, which encodes for the enzyme Tolueneo-Xylene Monooxygenase (ToMO) which can degrade the aromatic hydrocarbons from *Pseudomonas stutzeri*. It was cloned and expressed in Antarctic *Pseudoalteromonas* algae [88]. Genetically modified plants like *P. angustifolia*, *N. tabacum*, and *S. cucubalis* have been shown to accumulate more heavy metals pollutants than their wild counterparts by over expressing-glutamylcysteine-synthetase [132]. The γ-ECS *B. juncea* transgenic seedlings (*E. coli* gshl gene insertion) showed greater tolerance toward cadmium, phytochelatins, glutathione, and non-protein thiols than wild type [132]. The expression of genes including *AtNramps*, *AtPcrs*, *CAD1* (*A. thaliana*), *gshI*, *gshII* (*D. innoxia*), *CAX-2*, *NtCBP4* (*N. tabacum*; *A. thaliana*), *Ferretin* (*soybean*), *merA* (bacteria), and *PCS* cDNA clone (*B. juncea*) has been shown enhanced heavy metal tolerance and accumulation [133, 134]. However, in transgenic *A. thaliana*, the combined expression of SRSIp/ArsC and ACT 2p/γ-ECS presented increased tolerance to arsenic (Ar). Additionally, plants accrued 4 to 17-fold shoot biomass and accumulated 2 to 3-fold more AR compared to wild plants [135, 136]. For enhanced phytoremediation, metabolic pathways have been expressed by introducing *MerA, MerB, ars C,* and *γ-ECS* genes in *Arabidopsis* plants and resulted in enhanced accumulation of mercury, arsenate, and arsenite pollutants [137]. The over-expression of 1- amino cyclopropane-1-carboxylic acid deaminase in plants resulted in a higher accumulation of pollutants [75, 137]. Presently, the researchers aim to work on the genetic modification of common ornamental and houseplants for the improvement of indoor air quality. In this context, pothos ivy or devil's ivy has been modified by genetic engineering approach for the removal of chloroform and benzene by the expression of the Mammalian Cytochrome P450 *2e1* gene [138].

7. Conclusions

The removal or degradation of harmful pollutants from the air is far more challenging than water and soil pollution. As per the literature, phytoremediation has way more potential for eradication of harmful pollutants from the indoor and outdoor air environment in a sustainable manner. This way of phytoremediation is more sustainable, green, cost-effective, and easy to handle. In phytoremediation, the plants and microbes eradicate the pollutants through the following mechanisms, such as phyto-extraction, phyto-stabilization, rhizo-filtration, phyto-volatilization, phyto-degradation, and phyto-desalination. However, the selection of plant species and microbe's assistance for particular pollutant, their site, and condition are very crucial. As per the literature surveyed, in sustainability, the phytoremediation methods are 5–13 times more economical, higher success rate, good acceptance, and indeed an ecological way than the other remediation strategies. In the urban context, indoor plants serve as solution for phytoremediation against harmful air pollutants and also add esthetic value to the indoor space. Furthermore, the utilization of biotechnology will revolutionize and enhance the phytoremediation process against variety of pollutants.

Acknowledgements

The authors are grateful to the Director, CSIR-Institute of Himalayan Bioresource Technology (IHBT), Palampur, for providing all the necessary facilities. The authors would also like to acknowledge the Council of Scientific and Industrial Research (CSIR), Government of India, for providing the financial support under the project "CSIR- Floriculture Mission (HCP-0037)."

Funding

The study was funded by the Council of Scientific and Industrial Research (CSIR), Government of India, under the project "CSIR- Floriculture Mission" (Project Number: HCP-0037).

Disclosure statement

The authors declare that they have no known competing financial interests or personal relationships that could have appeared to influence the work reported in this paper.

Author details

Mahinder Partap[1,2†], Diksha Sharma[1†], Deekshith HN[1†], Anjali Chandel[1,2], Meenakshi Thakur[1], VipashaVerma[1] and Bhavya Bhargava[1*]

1 Floriculture Laboratory, Agrotechnology Division, Council of Scientific and Industrial Research (CSIR), Institute of Himalayan Bioresource Technology (IHBT), Palampur, HP, India

2 Academy of Scientific and Innovative Research (AcSIR), Ghaziabad, Uttar Pradesh, India

*Address all correspondence to: bhavya@ihbt.res.in

† Authors contributed equally to this work.

IntechOpen

© 2023 The Author(s). Licensee IntechOpen. This chapter is distributed under the terms of the Creative Commons Attribution License (http://creativecommons.org/licenses/by/3.0), which permits unrestricted use, distribution, and reproduction in any medium, provided the original work is properly cited.

References

[1] Manisalidis I, Stavropoulou E, Stavropoulos A, Bezirtzoglou E. Environmental and health impacts of air pollution: A review. Frontiers in public health. 2020;**8**:14

[2] Newby DE, Mannucci PM, Tell GS, Baccarelli AA, Brook RD, Donaldson K, et al. Expert position paper on air pollution and cardiovascular disease. European Heart Journal. 2015;**36**:83-93b

[3] Brook RD, Rajagopalan S, Pope CA III, Brook JR, Bhatnagar A, Diez-RouxAV, et al. Particulate matter air pollution and cardiovascular disease: Anupdate to the scientific statement from the American Heart Association.Circulation 2010;121:2331-2378.

[4] Chin MT. Basic mechanisms for adverse cardiovascularevents associated with air pollution. Heart. 2015;**101**:253-256. DOI: 10.1136/heartjnl-2014-306379

[5] Hamanaka RB, Mutlu GM. Particulate matter air pollution: Effects on the cardiovascular system. Frontiers in Endocrinology. 2018;**9**:680

[6] Bourdrel T, Bind MA, Béjot Y, Morel O, Argacha JF. Cardiovascular effects of air pollution. Archives of Cardiovascular Diseases. 2017;**110**(11):634-642

[7] Lelieveld J, Evans JS, Fnais M, Giannadaki D, Pozzer A. The contributionof outdoor air pollution sources to premature mortality on a global scale. Nature. 2015;**525**:367-371

[8] Brook RD, Newby DE, Rajagopalan S. The global threat of outdoor ambientair pollution to cardiovascular health: Time for intervention. JAMA Cardiology. 2017;**2**:353-354

[9] Zhang Y, Beggs PJ, McGushin A, Bambrick H, Trueck S, Hanigan IC, et al. The 2020 special report of the MJA–lancet countdown on health and climate change: Lessons learnt from Australia's "black summer". Medical Journal of Australia. 2020;**213**(11):490-492

[10] Weyens N, Thijs S, Popek R, Witters N, Przybysz A, Espenshade J, et al. The role of plant–microbe interactions and their exploitation for phytoremediation of air pollutants. International Journal of Molecular Sciences. 2015;**16**(10):25576-25604

[11] Landrigan PJ. Air pollution and health. The Lancet Public Health. 2017;**2**(1):4-5

[12] Bhargava B, Malhotra S, Chandel A, Rakwal A, Kashwap RR, Kumar S. Mitigation of indoor air pollutants using areca palm potted plants in real-life settings. Environmental Science and Pollution Research 2021. 2021;**28**(7):8898-8906

[13] Frontczak M, Andersen RV, Wargocki P. Questionnaire survey on factors influencing comfort with indoor environmental quality in Danish housing. Building and Environment. 2012;**50**:56-64

[14] Turner MC, Andersen ZJ, Baccarelli A, Diver WR, Gapstur SM, Pope CA III, et al. Outdoor air pollution and cancer: An overview of the current evidence and public health recommendations. CA: a Cancer Journal for Clinicians. 2020;**70**(6):460-479

[15] Ali H, Khan E, Sajad MA. Phytoremediation of heavy metals—Concepts and applications. Chemosphere. 2013;**91**(7):869-881

[16] Haritash AK, Kaushik CP. Biodegradation aspects of polycyclic aromatic hydrocarbons (PAHs): A review. Journal of Hazardous Materials. 2009;**169**(1-3):1-15

[17] Alagić SČ, Maluckov BS, Radojičić VB. How can plants manage polycyclic aromatic hydrocarbons? May these effects represent a useful tool for an effective soil remediation? A review. Clean Technologies and Environmental Policy. 2015;**17**:597-614

[18] Yang Y, Su Y, Zhao S. An efficient plant–microbe phytoremediation method to remove formaldehyde from air. Environmental Chemistry Letters. 2020;**18**(1):197-206

[19] Torpy F, Clements N, Pollinger M, Dengel A, Mulvihill I, He C, et al. Testing the single-pass VOC removal efficiency of an active green wall using methyl ethyl ketone (MEK). Air Quality, Atmosphere & Health. 2018;**11**(2):163-170

[20] Sly PD. Environmental pollutants and postnatal growth. In: Victor R, editor. Handbook of growth and growth monitoring in health and disease. Preedy. New York, USA: Springer; 2012. pp. 757-768. DOI: 10.1007/978-1-4419-1795-9_44

[21] Van Groenigen KJ, Osenberg CW, Hungate BA. Increased soil emissions of potent greenhouse gases under increased atmospheric CO_2. Nature. 2011;**475**(7355):214-216

[22] Tian ZH, Yang ZL. Scenarios of carbon emissions from the power sector in Guangdong province. Sustainability. 2016;**8**(9):863

[23] Azevedo VG, Sartori S, Campos LM. CO2 emissions: A quantitative analysis among the BRICS nations. Renewable and Sustainable Energy Reviews. 2018;**81**:107-115

[24] Freund P. Anthropogenic climate change and the role of CO2 capture and storage (CCS). In: Jon G, Mathais S, editors. Geological storage of carbon dioxide (CO2): Geoscience, technologies, environmental aspects and legal frameworks. UK: Woodhead publishing; 2013. pp. 1-23

[25] Bertoni G, Ciuchini C, Tappa R. Measurement of long-term average carbon dioxide concentrations using passive diffusion sampling. Atmospheric Environment. 2004;**38**(11):1625-1630

[26] Satish U, Mendell MJ, Shekhar K, Hotchi T, Sullivan D, Streufert S, et al. Is CO2 an indoor pollutant? Direct effects of low-to-moderate CO2 concentrations on human decision-making performance. Environmental Health Perspectives. 2012;**120**(12):1671-1677

[27] Law J, Watkins S, Alexander D. In-Flight Carbon Dioxide Exposures and Related Symptoms: Associations, Susceptibility and Operational Implications. NASA-STD-3001 Technical Brief. 2010. Available from: https://www.nasa.gov/sites/default/files/atoms/files/co2_technical_brief_ochmo.pdf

[28] Fierro MA, O'Rourke MK, Burgess JL. Adverse Health Effects of Exposure to Ambient Carbon Monoxide. University of Arizona Report; 2001. Available from: https://www.scribd.com/document/310600769/Adverse-Health-Effects-of-Exposure-to-Ambient-Carbon-Monoxide

[29] Prockop LD, Chichkova RI. Carbon monoxide intoxication: An updated review. Journal of the neurological sciences. 2007;**262**(1-2):122-130

[30] World Health Organization, WHO. Guidelines for drinking water quality. 2011. Available from: https://www.who.int/publications/i/item9789241549950 [Accessed: January 14, 2022]

[31] Goldstein NJ, Cialdini RB, Griskevicius V. A room with a viewpoint: Using social norms to motivate environmental conservation in hotels. Journal of consumer Research. 2008;**35**(3):472-482

[32] Lin B, Zhu J. Changes in urban air quality during urbanization in China. Journal of Cleaner Production. 2018;**188**:312-321

[33] Pandey P, Yadav R. A review on volatile organic compounds (VOCs) as environmental pollutants: Fate and distribution. International Journal of Plant and Environment. 2018;**4**(02):14-26

[34] Campagnolo D, Saraga DE, Cattaneo A, Spinazzè A, Mandin C, Mabilia R, et al. VOCs and aldehydes source identification in European office buildings-the OFFICAIR study. Building and Environment. 2017;**115**:18-24

[35] Gonzalez-Martin J, Kraakman NJ, Perez C, Lebrero R, Munoz R. A state–of–the-art review on indoor air pollution and strategies for indoor air pollution control. Chemosphere. 2021;**262**:128376

[36] Cheung K, Daher N, Kam W, Shafer MM, Ning Z, Schauer JJ, et al. Spatial and temporal variation of chemical composition and mass closure of ambient coarse particulate matter (PM10-2.5) in the Los Angeles area. Atmospheric environment. 2011;**45**(16):2651-2662

[37] Yoon HI, Hong YC, Cho SH, Kim H, Kim YH, Sohn JR, et al. Exposure to volatile organic compounds and loss of pulmonary function in the elderly. European Respiratory Journal. 2013;**6**(6):1270-1276

[38] Ukaogo PO, Ewuzie U, Onwuka CV. Environmental pollution: Causes, effects, and the remedies. In: Microorganisms for Sustainable Environment and Health. Elsevier; 2020. pp. 419-429. DOI: 10.1016/C2018-0-05025-2

[39] Sánchez-Soberón F, Rovira J, Mari M, Sierra J, Nadal M, Domingo JL, et al. Main components and human health risks assessment of PM10, PM2. 5, and PM1 in two areas influenced by cement plants. Atmospheric Environment. 2015;**120**:109-116

[40] Priyamvada H, Priyanka C, Singh RK, Akila M, Ravikrishna R, Gunthe SS. Assessment of PM and bioaerosols at diverse indoor environments in a southern tropical Indian region. Building and Environment. 2018;**137**:215-225

[41] Smith SB, Gill CA, Lunt DK, Brooks MA. Regulation of fat and fatty acid composition in beef cattle. Asian-Australasian Journal of Animal Sciences. 2009;**22**(9):1225-1233

[42] Boogaard BK, Bock BB, Oosting SJ, Wiskerke JS, van der Zijpp AJ. Social acceptance of dairy farming: The ambivalence between the two faces of modernity. Journal of Agricultural and Environmental Ethics. 2011;**24**:259-282

[43] Kiurski JS, Marić BB, Aksentijević SM, Oros IB, Kecić VS, Kovacˇević IM. Indoor air quality investigation from screen printing industry. Renewable and Sustainable Energy Reviews. 2013;**28**:224-231

[44] Abdel-Shafy HI, Mansour MS. A review on polycyclic aromatic hydrocarbons: Source, environmental impact, effect on human health and remediation. Egyptian Journal of Petroleum. 2016;**25**(1):107-123

[45] Pongpiachan S, Tipmanee D, Khumsup C, Kittikoon I, Hirunyatrakul P. Assessing risks to adults and preschool

children posed by PM2. 5-bound polycyclic aromatic hydrocarbons (PAHs) during a biomass burning episode in northern Thailand. Science of the Total Environment. 2015;**508**:435-444

[46] Domingo JL, Nadal M. Human dietary exposure to polycyclic aromatic hydrocarbons: A review of the scientific literature. Food and Chemical Toxicology. 2015;**86**:144-153

[47] Akyüz M, Çabuk H. Meteorological variations of PM2. 5/PM10 concentrations and particle-associated polycyclic aromatic hydrocarbons in the atmospheric environment of Zonguldak, Turkey. Journal of Hazardous Materials. 2009;**170**(1):13-21

[48] Lannerö E, Wickman M, van Hage M, Bergström A, Pershagen G, Nordvall L. Exposure to environmental tobacco smoke and sensitisation in children. Thorax. 2008;**63**(2):172-176

[49] Kim KH, Jahan SA, Kabir E, Brown RJ. A review of airborne polycyclic aromatic hydrocarbons (PAHs) and their human health effects. Environment International. 2013;**60**:71-80

[50] Hamid N, Syed JH, Junaid M, Mahmood A, Li J, Zhang G, et al. Elucidating the urban levels, sources and health risks of polycyclic aromatic hydrocarbons (PAHs) in Pakistan: Implications for changing energy demand. Science of the Total Environment. 2018;**619**:165-175

[51] Gryparis A, Forsberg B, Katsouyanni K, Analitis A, Touloumi G, Schwartz J, et al. Acute effects of ozone on mortality from the "air pollution and health: A European approach" project. American journal of respiratory and critical care medicine. 2004;**170**(10):1080-1087

[52] Siegel RL, Miller KD, Wagle NS, Jemal A. Cancer statistics, 2023. CA: a cancer journal for clinicians. 2023;**73**(1):17-48

[53] Rodríguez D, Cobo-Cuenca AI, Quiles R. Effects of air pollution on daily hospital admissions for cardiovascular diseases in Castilla-La Mancha, Spain: A region with moderate air quality. Air Quality, Atmosphere & Health. 2022;**15**(4):591-604

[54] Gu X, Wang T, Li C. Elevated ozone decreases the multifunctionality of belowground ecosystems. Global Change Biology. 2023;**29**(3):890-908

[55] Gawande P, Kaware J. Health and environmental effects of Sulphur oxides—A review. International Journal of Science and Research. 2015;**6**(6):1262-1265

[56] Jafarinejad S. Control and Treatment of Air Emissions. In: Jafarinejad S, Butterworth-Heinemann, editors. Petroleum Waste Treatment and Pollution Control. 2017. Pages 149-183. ISBN 9780128092439, DOI: 10.1016/B978-0-12-809243-9.00005-5. Available from: https://www.sciencedirect.com/science/article/pii/B9780128092439000055

[57] Chen C, Arjomandi M, Balmes J, Tager I, Holland N. Effects of chronic and acute ozone exposure on lipid peroxidation and antioxidant capacity in healthy young adults. Environmental health perspectives. 2007;**115**(12):1732-1737

[58] Shabala S. Regulation of potassium transport in leaves: From molecular to tissue level. Annals of Botany. 2003;**92**(5):627-634

[59] Demidchik V, Shabala SN, Coutts KB, Tester MA, Davies JM. Free oxygen

radicals regulate plasma membrane Ca^{2+}-and K^{+}-permeable channels in plant root cells. Journal of cell science. 2003;**116**(1):81-88

[60] Gawronski SW, Gawronska H, Lomnicki S, Saebo A, Vangronsveld J. Plants in air phytoremediation. In: Advances in Botanical Research. Vol. 83. United states, Elseviar; 2017. pp. 319-346. DOI: 10.1016/bs.abr.2016.12.008

[61] Agarwal A, Poelchau MH, Kenkmann T, Rae A, Ebert M. Impact experiment on gneiss: The effects of foliation on cratering process. Journal of Geophysical Research: Solid Earth. 2019;**124**(12):13532-13546

[62] Teiri H, Hajizadeh Y, Azhdarpoor A. A review of different phytoremediation methods and critical factors for purification of common indoor air pollutants: An approach with sensitive analysis. Air Quality, Atmosphere & Health. 2021;**15**:1-19

[63] Ciffroy P, Tanaka T. Modelling the fate of Chemicals in Plants. In: Ciffroy P, Tediosi A, Capri E, editors. Modelling the Fate of Chemicals in the Environment and the Human Body. Switzerland: Springer Cham; 2018. pp. 167-189. DOI: 10.1007/978-3-319-59502-3_8

[64] Mohan I, Goria K, Dhar S, Kothari R, Bhau BS, Pathania D. Phytoremediation of heavy metals from the biosphere perspective and solutions. In: Pollutants and Water Management: Resources, Strategies and Scarcity. USA: Wiley-Blackwell; 2021. pp. 95-127. DOI: 10.1002/9781119693635.ch5

[65] Winter K, Holtum JA. Facultative crassulacean acid metabolism (CAM) plants: Powerful tools for unravelling the functional elements of CAM photosynthesis. Journal of experimental botany. 2014;**65**(13):3425-3441

[66] Waghmode M, Gunjal A, Patil N, Shinde S. Composition and interconnections in Phyllomicrobiome. In: Phytomicrobiome Interactions and Sustainable Agriculture. USA: Wiley-Blackwell; 2021. pp. 277-292. DOI: 10.1002/9781119644798.ch15

[67] Molina L, Wittich RM, van Dillewijn P, Segura A. Plant-bacteria interactions for the elimination of atmospheric contaminants in cities. Agronomy. 2021;**11**(3):493

[68] Chen L, Yang JY, Wang D. Phytoremediation of uranium and cadmium contaminated soils by sunflower (Helianthus annuus L.) enhanced with biodegradable chelating agents. Journal of Cleaner Production. 2020;**263**:121491

[69] Gul I, Manzoor M, Hashmi I, Bhatti MF, Kallerhoff J, Arshad M. Plant uptake and leaching potential upon application of amendments in soils spiked with heavy metals (Cd and Pb). Journal of environmental management. 2019;**249**:109408

[70] Mousavi A, Pourakbar L, Moghaddam SS, Popović-Djordjević J. The effect of the exogenous application of EDTA and maleic acid on tolerance, phenolic compounds, and cadmium phytoremediation by okra (*Abelmoschus esculentus* L.) exposed to Cd stress. Journal of Environmental Chemical Engineering. 2021;**9**(4):105456

[71] Morikawa H, Erkin ÖC. Basic processes in phytoremediation and some applications to air pollution control. Chemosphere. 2003;**52**(9):1553-1558

[72] Krishnasamy S, Lakshmanan R, Ravichandran M. Phytoremediatiation of metal and metalloid pollutants from farmland: An In-situ soil conservation. In: Biodegradation Technology of

Organic and Inorganic Pollutants. London, UK: IntechOpen; 2021. DOI: 10.5772/intechopen.98659

[73] Favas PJ, Pratas J, Varun M, D'Souza R, Paul MS. Phytoremediation of soils contaminated with metals and metalloids at mining areas: Potential of native flora. Environmental risk assessment of soil contamination. 2014;**3**:485-516

[74] KhKaraghool HA. The employment of Endophytic bacteria forPhytodegradation of pyridine. IOP Conference Series Earth and Environmental Science. 2021;**961**:961

[75] Kumar K, Shinde A, Aeron V, Verma A, Arif NS. Genetic engineering of plants for phytoremediation: Advances and challenges. Journal of Plant Biochemistry and Biotechnology. 2022;**32**:1-19

[76] Beckett KP, Freer-Smith PH, Taylor G. Particulate pollution capture by urban trees: Effect of species and windspeed. Global change biology. 2000;**6**(8):995-1003

[77] Roupsard P, Amielh M, Maro D, Coppalle A, Branger H, Connan O, et al. Measurement in a wind tunnel of dry deposition velocities of sub micron aerosol with associated turbulence onto rough and smooth urban surfaces. Journal of Aerosol Science. 2013;**55**:12-24

[78] Treesubsuntorn C, Thiravetyan P. Botanical biofilter for indoor toluene removal and reduction of carbon dioxide emission under low light intensity by using mixed C3 and CAM plants. Journal of Cleaner Production. 2018;**194**:94-100

[79] Ny MT, Lee BK. Size distribution of airborne particulate matter and associated metallic elements in an urban area of an industrial city in Korea. Aerosol and Air Quality Research. 2011;**11**(6):643-653

[80] Mate AR, Deshmukh RR. To control effects of air pollution using roadside trees. International Journal of Innovative Research in Science, Engineering and Technology. 2015;**4**(11):11167-11172

[81] Wolffe MC, Wild O, Long SP, Ashworth K. Temporal variability in the impacts of particulate matter on crop yields on the North China plain. Science of the Total Environment. 2021;**776**:145135

[82] Wu W, Huang ZH, Lim TT. Recent development of mixed metal oxide anodes for electrochemical oxidation of organic pollutants in water. Applied Catalysis A: General. 2014;**480**:58-78

[83] Vejerano EP, Rao G, Khachatryan L, Cormier SA, Lomnicki S. Environmentally persistent free radicals: Insights on a new class of pollutants. Environmental science & technology. 2018;**52**(5):2468-2481

[84] Kelly FJ, Fussell JC. Air pollution and public health: Emerging hazards and improved understanding of risk. Environmental geochemistry and health. 2015;**37**:631-649

[85] Kiruri LW, Khachatryan L, Dellinger B, Lomnicki S. Effect of copper oxide concentration on the formation and persistency of environmentally persistent free radicals (EPFRs) in particulates. Environmental science & technology. 2014;**48**(4):2212-2217

[86] Pandey VC, Bajpai O. Chapter 1 - phytoremediation: From theory toward practice. In: Pandey C, Bauddh K, editors. Phytomanagement of Polluted Sites. Vimal CP, Kuldeep B. editors. United states: Elsevier; 2019. pp. 1-49. ISBN 9780128139127,

DOI: 10.1016/B978-0-12-813912-7.00026-0. Available from: https://www.sciencedirect.com/science/article/pii/B9780128139127000260. DOI: 10.1016/C2017-0-00586-4

[87] Agarwal P, Sarkar M, Chakraborty B, Banerjee T. Phytoremediation of air pollutants: Prospects and challenges. In: Vimal CP, Kuldeep B, editors. Phytomanagement of Polluted Sites. United states: Elsevier; 2019. pp. 221-241. ISBN 9780128139127

[88] Doty SL, Andrew JC, Moore AL, Vajzovic A, Singleton GL, Ma CP. Enhanced phytoremediation of volatile environmental pollutants with transgenic trees. Proceedings. National Academy of Sciences. United States of America. 2007;**104**(43):1681616821

[89] Engel-Di Mauro S. Atmospheric sources of trace element contamination in cultivated urban areas: A review. Journal of Environmental Quality. 2021;**50**(1):38-48

[90] Farr C. Indoor Air Quality: Causes, Controls and Consequences. United Kingdom: Lancaster University; 2021

[91] Toome M, Randjärv P, Copolovici L, Niinemets Ü, Heinsoo K, Luik A, et al. Leaf rust induced volatile organic compounds signaling in willow during the infection. Planta. 2010;**232**:235-243

[92] Mikkonen A, Li T, Vesala M, Saarenheimo J, Ahonen V, Kärenlampi S, et al. Biofiltration of airborne VOC s with green wall systems—Microbial and chemical dynamics. Indoor Air. 2018;**28**(5):697-707

[93] Lomonaco T, Manco E, Corti A, La Nasa J, Ghimenti S, Biagini D, et al. Release of harmful volatile organic compounds (VOCs) from photo-degraded plastic debris: A neglected source of environmental pollution. Journal of hazardous materials. 2020;**394**:122596

[94] Barac T, Taghavi S, Borremans B, Provoost A, Oeyen L, Colpaert JV, et al. Engineered endophytic bacteria improve phytoremediation of water-soluble, volatile, organic pollutants. Nature biotechnology. 2004;**22**(5):583-588

[95] Sandhu A, Halverson LJ, Beattie GA. Identification and genetic characterization of phenol-degrading bacteria from leaf microbial communities. Microbial ecology. 2009;**57**:276-285

[96] De Kempeneer L, Sercu B, Vanbrabant W, Van Langenhove H, Verstraete W. Bioaugmentation of the phyllosphere for the removal of toluene from indoor air. Applied microbiology and biotechnology. 2004;**64**:284-288

[97] Sriprapat W, Roytrakul S, Thiravetyan P. Proteomic studies of plant and bacteria interactions during benzene remediation. Journal of Environmental Sciences. 2020;**94**:161-170

[98] Zhang X, Wang H, He L, Lu K, Sarmah A, Li J, et al. Using biochar for remediation of soils contaminated with heavy metals and organic pollutants. Environmental Science and Pollution Research. 2013;**20**:8472-8483

[99] Wang S, Xing J, Zhao B, Jang C, Hao J. Effectiveness of national air pollution control policies on the air quality in metropolitan areas of China. Journal of Environmental Sciences. 2014;**26**(1):13-22

[100] Asif M, Saleem S, Tariq A, Usman M, Haq RAU. Pollutant emissions from brick kilns and their effects on climate change and agriculture. ASEAN Journal of Science and Engineering. 2021;**1**(2):135-140

[101] Kim KJ, Kim HJ, Khalekuzzaman M, Yoo EH, Jung HH, Jang HS. Removal ratio of gaseous toluene and xylene transported from air to root zone via the stem by indoor plants. Environmental Science and Pollution Research. 2016;**23**:6149-6158

[102] Roupsard P, Amielh M, Maro D, Coppalle A, Branger H, Connan O, et al. Measurement in a wind tunnel of dry deposition velocities of submicron aerosol with associated turbulence onto rough and smooth urban surfaces. Journal of Aerosol Science. 2013;**55**:12-24

[103] Hu H, Zhou Q, Li X, Lou W, Du C, Teng Q, et al. Phytoremediation of anaerobically digested swine wastewater contaminated by oxytetracycline via Lemna aequinoctialis: Nutrient removal, growth characteristics and degradation pathways. Bioresource technology. 2019;**291**:121853

[104] Yang L, Wang J, Yang Y, Li S, Wang T, Oleksak P, et al. Phytoremediation of heavy metal pollution: Hotspots and future prospects. Ecotoxicology and Environmental Safety. 2022;**234**:113403

[105] Ho CP, Hseu ZY, Chen NC, Tsai CC. Evaluating heavy metal concentration of plants on a serpentine site for phytoremediation applications. Environmental earth sciences. 2013;**70**:191-199

[106] Verma VK, Singh YP, Rai JP. Biogas production from plant biomass used for phytoremediation of industrial wastes. Bioresource technology. 2007;**98**(8):1664-1669

[107] Zhang H, Cai X, Song Y, Bian Y, Xiao K, Zhang H. Influence of intermittent turbulence on air pollution and its dispersion in winter 2016/2017 over Beijing, China. Journal of Meteorological Research. 2020;**34**(1):176-188

[108] Li Y, Liu B, Xue Z, Zhang Y, Sun X, Song C, et al. Chemical characteristics and source apportionment of PM2. 5 using PMF modelling coupled with 1-hr resolution online air pollutant dataset for Linfen, China. Environmental Pollution. 2020;**263**:114532

[109] Yutthammo C, Thongthammachat N, Pinphanichakarn P, Luepromchai E. Diversity and activity of PAH-degrading bacteria in the phyllosphere of ornamental plants. Microbial ecology. 2010;**59**:357-368

[110] Wang Y, Fang L, Lin L, Luan T, Tam NF. Effects of low molecular-weight organic acids and dehydrogenase activity in rhizosphere sediments of mangrove plants on phytoremediation of polycyclic aromatic hydrocarbons. Chemosphere. 2014;**99**:152-159

[111] Li J, Zhang J, Larson SL, Ballard JH, Guo K, Arslan Z, et al. Electrokinetic-enhanced phytoremediation of uranium-contaminated soil using sunflower and Indian mustard. International Journal of Phytoremediation. 2019;**21**(12):1197-1204

[112] Alagić SČ, Jovanović VP, Mitić VD, Cvetković JS, Petrović GM, Stojanović GS. Bioaccumulation of HMW PAHs in the roots of wild blackberry from the Bor region (Serbia): Phytoremediation and biomonitoring aspects. Science of the Total Environment. 2016;**562**:561-570

[113] Glick BR. Using soil bacteria to facilitate phytoremediation. Biotechnology advances. 2010;**28**(3): 367-374

[114] Lin Q, Mendelssohn IA. Determining tolerance limits for restoration and

phytoremediation with Spartina patens in crude oil-contaminated sediment in greenhouse. Archives of Agronomy and Soil Science. 2008;**54**(6):681-90.4

[115] Shao B, Liu Z, Zhong H, Zeng G, Liu G, Yu M, et al. Effects of rhamnolipids on microorganism characteristics and applications in composting: A review. Microbiological Research. 2017;**200**:33-44

[116] Liu G, Xiao M, Zhang X, Gal C, Chen X, Liu L, et al. A review of air filtration technologies for sustainable and healthy building ventilation. Sustainable Cities and Society. 2017;**32**:375-396

[117] Liu H, Luo L, Jiang G, Li G, Zhu C, Meng W, et al. Sulfur enhances cadmium bioaccumulation in Cichorium intybus by altering soil properties, heavy metal availability and microbial community in contaminated alkaline soil. Science of The Total Environment. 2022;**837**:155879

[118] Irga PJ, Paull NJ, Abdo P, Torpy FR. An assessment of the atmospheric particle removal efficiency of an in-room botanical biofilter system. Building and Environment. 2017;**115**:281-290

[119] Bandehali S, Parvizian F, Hosseini SM, Matsuura T, Drioli E, Shen J, et al. Planning of smart gating membranes for water treatment. Chemosphere. 2021;**283**:131207

[120] Ogut O, Tzortzi NJ, Bertolin C. Vertical green structures to establish sustainable built environment: A systematic market review. Sustainability. 2022;**14**(19):12349

[121] Tomson M, Kumar P, Barwise Y, Perez P, Forehead H, French K, et al. Green infrastructure for air quality improvement in street canyons. Environment International. 2021;**146**:106288

[122] Moya TA, Dobbelsteen AVD, Ottelé M, Bluyssen PM. A review of green systems within the indoor environment. Indoor and Built Environment. 2018;**28**:298-230

[123] Modirrousta S, Mohammadi Z. Necessity and methods of designing green buildings in cities and its effect on energy efficiency. European Online Journal of Natural and Social Sciences Proceedings. 2015;**4**:304-314

[124] Khalifa AA, Khan E, Akhtar MS. Phytoremediation of indoor formaldehyde by plants and plant material. International Journal of Phytoremediation. 2022;**25**:1-12

[125] Prakash G, Soni R, Mishra R, Sharma S. Role of plant-microbe interaction in phytoremediation. In: In Vitro Plant Breeding towards Novel Agronomic Traits: Biotic and Abiotic Stress Tolerance. Switzerland: Springer Cham; 2019. pp. 83-118. DOI: 10.1007/978-981-32-9824-8_6

[126] Vafaie Moghadam A, Iranbakhsh A, Saadatmand S, Ebadi M, Oraghi AZ. New insights into the transcriptional, epigenetic, and physiological responses to zinc oxide nanoparticles in Datura stramonium; potential species for phytoremediation. Journal of Plant Growth Regulation. 2021;**41**:271-281

[127] Kazemi M. Screening of native plant species for phytoremediation potential in Pb-Zn mines in Iran. Environmental Resources Research. 2021;**9**(1):89-98

[128] Fleck R, Gill RL, Pettit T, Irga PJ, Williams NL, Seymour JR, et al. Characterisation of fungal and bacterial dynamics in an active green wall used for indoor air pollutant removal. Building and Environment. 2020;**179**:106987

[129] Kurade MB, Ha YH, Xiong JQ, Govindwar SP, Jang M, Jeon BH.

Phytoremediation as a green biotechnology tool for emerging environmental pollution: A step forward towards sustainable rehabilitation of the environment. Chemical Engineering Journal. 2021;**415**:129040

[130] Rai A, Yamazaki M, Saito K. A new era in plant functional genomics. Current Opinion in Systems Biology. 2019;**15**:58-67

[131] Jonathan A. Vegetation Climate Interaction: How Vegetation Makes the Global Environment. Switzerland: Springer Cham; 2003. DOI: 10.1007/978-3-642-00881-8

[132] James CA, Strand SE. Phytoremediation of small organic contaminants using transgenic plants. Current Opinion in Biotechnology. Switzerland: Springer Cham; 2009;**20**:237-241. DOI: 10.1007/978-3-642-00881-8

[133] Sana B. Bio resources for control of environmental pollution. In: Biotechnological Applications of Biodiversity. Berlin, Heidelberg: Springer; 2014. pp. 137-183

[134] Miri S, Naghdi M, Rouissi T, Kaur BS, Martel R. Recent biotechnological advances in petroleum hydrocarbons degradation under cold climate conditions: A review. Critical Reviews in Environmental Science and Technology. 2019;**49**(7):553-586

[135] Adams GO, Fufeyin PT, Okoro SE, Ehinomen I. Bioremediation, biostimulation and bioaugmention: A review. International Journal of Environmental Bioremediation & Biodegradation. 2015;**3**(1):28-39

[136] Glick BR, Patten CL. Molecular Biotechnology: Principles and Applications of Recombinant DNA. USA: Wiley-Blackwell; 2022. ISBN: 978-1-683-67366-8. Available from: https://www.wiley.com/en-us/Molecular+Biotechnology%3A+Principles+and+Applications+of+Recombinant+DNA%2C+6th+Edition-p-9781683673668

[137] Siani L, Papa R, Di Donato A, Sannia G. Recombinant expression of Tolueneo-xylene monooxygenase (ToMO) from *pseudomonas stutzeri* OX1 in the marine *Antarctic bacterium Pseudoalteromonas haloplanktis* TAC125. Journal of Biotechnology. 2006;**126**(3):334-341

[138] Erkin OC, Takahashi M, Morikawa H. Development of a regeneration and transformation system for *Raphiolepis umbellata* L., "Sharinbai" plants by using particle bombardment. Plant Biotechnology. 2003;**20**:145-152

Chapter 4

Fine Particles in the Ambient Air as a Risk Factor of Bronchial Asthma in Adults

Liliya M. Fatkhutdinova, Olesya V. Skorohodkina, Laila I. Yapparova, Guzel A. Timerbulatova and Ramil R. Zalyalov

Abstract

Air pollution with suspended particles and gaseous substances is assumed to be a possible risk factor for bronchial asthma. Bronchial asthma (BA) is one of the most common chronic non-communicable diseases in children and adults, characterized by variable respiratory symptoms and airflow limitation. Asthma is a heterogeneous disease with different underlying disease processes. The most common asthma phenotypes are allergic and non-allergic asthma, differing in the presence of atopy, the type of airway inflammation, responses to inhaled corticosteroid treatment. Meta-analyses, including cohort studies, support the role of fine particles in asthma in children. The question of whether the incidence of asthma in adults is associated with exposure to ambient particulate matter remains open. The chapter describes the effect of fine particles in the ambient air on the formation, course, and underlying mechanisms of different phenotype of bronchial asthma in adults. The role of ambient fine particles in the development of the eosinophilic non-allergic phenotype of bronchial asthma in adults (18–65 years old) has been proven. The hypothesis about different underlying mechanisms in response to exposure to particulate matter for various phenotypes of bronchial asthma was confirmed.

Keywords: bronchial asthma, asthma phenotypes, T2-endotype, fine particles, ambient air, epidemiological studies, adults

1. Introduction

Epidemiological studies from around the world indicate that particulate matter poses a serious health threat to public health [1]. Air pollution with suspended particles and gaseous substances is assumed to be a possible risk factor for bronchial asthma [2–4]. Bronchial asthma (BA) is one of the most common chronic non-communicable diseases in children and adults, characterized by variable respiratory symptoms and airflow limitation [5]. The global prevalence of physician-diagnosed asthma in adults is 4.3% (95% CI: 4.2%, 4.4%), with large differences between

 IntechOpen

countries [6]. Asthma is a heterogeneous disease with different underlying disease processes. Recognizable clusters of demographic, clinical, and/or pathophysiological characteristics are often called as asthma phenotypes [4]. Some of the most common are allergic and non-allergic asthma, differing in the presence of atopy, the type of airway inflammation, responses to inhaled corticosteroid treatment. Later, the term "endotype" was introduced as a conceptual basis for new ideas about the molecular heterogeneity of bronchial asthma [7], and T2- and non-T2-endotypes have now been described. Meta-analyses, including cohort studies, support the role of fine particles in asthma in children [8–11]. The question of whether the incidence of asthma in adults is associated with exposure to ambient particulate matter (PM) remains open: there are not enough studies, and the available data are contradictory—the relative risks were about 1.0 and, in most studies, did not reach a critical level of statistical significance [12–17]. Besides, adult asthma, unlike asthma in children, is associated with other risk factors and is known for its female predominance, uncommon remission, and unusual mortality [18]. The chapter describes the current literature data, as well as our study on the effect of fine particles in the ambient air on the formation, course, and possible underlying mechanisms of atopic allergic and eosinophilic non-allergic phenotypes of the T2-endotype of bronchial asthma in adults (18–65 years old).

Bronchial asthma is a heterogeneous disease characterized by the chronic inflammation of the airways [6]. The known variants of the combination of demographic, clinical, and/or pathophysiological characteristics are often called "bronchial asthma phenotypes" [6]. Several studies have shown that air pollution with PM increases the risk of bronchial asthma exacerbations and frequency of hospitalizations [2, 3, 19, 20] and worsens the quality of life of patients with asthma [21]. Meanwhile, the role of PM in the onset of bronchial asthma in adults is still open [22].

The earliest study reporting an association between long-term exposure to air pollution and the incidence of bronchial asthma described a cohort of non-smoking Seventh-day Adventists in California, USA [12]. Considering gender, age, education, smoking, and gaseous pollutants (ozone and sulfur dioxide) as confounders, no association was found between new cases of bronchial asthma and PM10 in the ambient air.

A Swiss cohort study with an 11-year follow-up showed that the incidence of bronchial asthma among non-smokers was associated with an increase in PM10 concentrations: hazard ratio 1.30 (95% CI: 1.05, 1.61) per 1 $\mu g/m^3$ of PM10, not being changed when adjusted by education, occupational exposure, secondhand smoking, asthma or allergies in parents, exposure to other pollutants, proximity to roads with heavy traffic, and functional state of the lungs [13].

A meta-analysis of the incidence of bronchial asthma among the adult population in six prospective cohorts followed up within the ESCAPE study revealed a positive but insignificant relationship between new cases of bronchial asthma and average annual concentrations of PM10 and PM2.5: odds ratio 1.04 (95% CI: 0.88, 1.23) and 1.04 (95% CI: 0.88, 1.23) by 10 $\mu g/m^3$ PM10 and 5 $\mu g/m^3$ PM2.5, respectively. The models included PM concentrations, as well as such confounders as gender, age, education, body mass index, smoking, and clinical aspects of BA [14]. Similar results were obtained in a cohort of women living in the USA (follow-up period 2008–2012), where the odds ratio for new cases of bronchial asthma, adjusted by age, education, body mass index, consumption of dietary fiber, smoking, and occupational hazards, was 1.20 (95% CI: 0.99, 1.46) for an increase in PM2.5 concentrations by 3.6 $\mu g/m^3$ (interquartile range) [15].

In older age groups, a 2-year increase in PM2.5 by 10 $\mu g/m^3$ was associated with an increased risk of bronchial asthma by 2.24% (95% CI: 0.93%, 5.38%) in people aged >44 years [16], a 3-year increase in the average annual concentration of PM2.5 by 10 $\mu g/m^3$ led to an increase in the bronchial asthma incidence by 9% (95% CI: 4%, 14%) among elderly people (65+) [17]. Similar results were reported by [23] for low- and middle-income countries (China, India, Ghana, Mexico, Russia, and South Africa): 5.12% of the asthma cases in the study population over 50 years of age (95% CI: 1.44%, 9.23%) could be attributed to long-term exposure to PM2.5.

The phenotypic heterogeneity of bronchial asthma was investigated by a cluster analysis of well-characterized patients, grouping them into 4–5 phenotypic clusters, considering age, gender, lung function, medical aid need, and body mass index [24, 25]. To date, the T2 and non-T2 asthma endotypes, defined as "a disease subtype that is determined by a separate functional or pathological biological mechanism" [7], have been described. T2-endotype is characterized by a high level of type 2 inflammatory response in the airways [3, 7] and more severe course [6]. The molecular mechanisms of the non-T2-endotype are under investigation [3, 7, 26]. Considering different asthma phenotypes and endotypes when studying health effects of ambient particles was not regarded in previous epidemiological studies, but several authors hypothesized that the bronchial asthma linked to PM might be described as a separate phenotype, and its initial mechanisms could include damage to the airway epithelium, T2- as well as T17-mediated responses [27, 28].

The issue of the relationship between bronchial asthma and separate fractions of PM in the ambient air is insufficiently studied [2, 3]. There are also no convincing data on the effects of ambient particles with different chemical composition or origin. There is some information about the role of the oxidizing potential, which depends on the chemical composition of suspended particles. As discussed above, reactive oxygen species induced by particulate matter are regarded as an important mediator of their toxicity. The oxidation potential of PM2.5 taken from the atmospheric air of Paris was increased in the presence of metals such as copper and zinc, as well as polycyclic aromatic hydrocarbons and soluble organic compounds in PM [29]. The effects of fine particulate matter were enhanced by concomitant exposure to particulate matter and bacterial endotoxins in residential air: for emergency medical visits due to asthma exacerbations in the last 12 months, the odds ratio for comparing the subgroup with high exposures to PM2.5 and endotoxin and the subgroup with low levels of both pollutants was 5.01 (95% CI: 2.54, 9.87) [30]. Similar findings were shown in a recent Japanese study [31]. Combined exposure to PM10 and bacterial endotoxin near livestock farms was associated with the higher prevalence of bronchial asthma [32]. Thus, this line of research also deserves attention.

2. Material and methods

The research aim was to study the effect of fine particles in the ambient air on the formation, course, and underlying mechanisms of atopic allergic and eosinophilic non-allergic phenotypes of the T2-endotype of bronchial asthma in adults (18–65 years old).

The study included the following parts: (1) a case-control study, (2) a biomarker study as a part of the case-control study, (3) an epidemiological study of ecological type based on geoinformatics approach with a retrospective analysis of data on environmental pollution and population health.

For the case-control study, patients with bronchial asthma were selected while seeking medical help ("cases"), and the comparison group was selected from among those who did not suffer from bronchial asthma ("controls"). The groups were formed based on inclusion/exclusion criteria and comparison criteria, supplemented by the collection of information about potential confounders. The probable T2-endotype of bronchial asthma was determined by the absolute number of eosinophils in the blood (≥150 cells/μl). A total of 156 patients with bronchial asthma were examined, of which 82 patients were selected in the "cases" group (40 patients with an allergic phenotype, 42 patients with an eosinophilic non-allergic phenotype of bronchial asthma). The inclusion criteria for the "cases" group were: (1) age from 18 to 65 years; (2) an established clinical diagnosis of an allergic or non-allergic phenotype of the T2-endotype of bronchial asthma; (3) informed consent to participate in the study. The exclusion criterion for the group of "cases" was the allergen-specific immunotherapy or biological therapy at the time of examination, or information in the medical records about the use of such therapy earlier. The comparison group (48 people) was selected according to the following comparison criteria: (1) compliance of the distribution of "controls" with the distribution of "cases" by sex, age (in the range up to 10 years), body mass index (up to 23.9; 24–29.9; 30 and more kg/m^2), level of education (secondary; college; high); (2) exclusion of the diagnosis of bronchial asthma and other chronic respiratory diseases; (3) informed consent to participate in the study. For individuals included in the study, the average and maximal annual concentrations of PM2.5 and PM10 fractions averaged over the period 2014–2020 were determined, considering measurements at monitoring points closest to the areas of residence. For measurements, the DustTrak™ II Aerosol Monitor 8530 (TSI Inc., USA) was used. Additionally, in the areas of residence, ambient air sampling was carried out by the 8-stage impactor MOUDI 100NR (TSI, USA) to study the elemental composition of the aerosol (SEM, energy dispersive spectroscopy) and the contamination by bacterial endotoxin (kinetic LAL test). Besides, air samples were taken for microbiological examination by classical cultural methods and using MALDI-TOF spectrometry. Using the multiple logistic regression, adjusted odds ratios were calculated with 95% confidence intervals for allergic and eosinophilic non-allergic phenotypes of bronchial asthma in comparison with the comparison group, depending on the levels of exposure variables characterizing air pollution with particulate matter.

For the biomarker study, as a part of the case-control study, 61 patients with T2-endotype of BA were examined (34 patients with an allergic phenotype, 27 patients with a non-allergic phenotype of the disease). The comparison group consisted of 30 people without symptoms of asthma and other allergic diseases who were matched by gender, age, body mass index (BMI), profession (position). All patients with BA and persons from the comparison group underwent blood sampling to determine biological markers of various types of inflammation: alarmins (TSLP, IL-33, IL-25), T2-cytokines (IL-4, IL-5, IL-13), DPP4, and also—in order to clarify the involvement of non-T2-mechanisms—IL-6, TGF-beta1, IL-17A, IL-1beta; multiplex analysis using xMAP Luminex technology was applied. The blood serum level of periostin was determined by ELISA. The study design was supplemented by whole blood sampling from the same study participants and subsequent analysis of the expression of genes encoding certain cytokines: IL-4, IL-5, IL-6, TGF-beta1, IL-17A, IL-1beta, IL-25, IL-33. In the biomarker study, the calculated masses of aerosol particles deposited in different parts of the lungs were used as additional exposure characteristics. To estimate the masses of aerosol particles deposited in the lungs, an original method to reconstruct the aerosol particle size distribution function using actual PM2.5 and

PM10 concentrations under the assumption of a lognormal distribution characteristic of atmospheric aerosols was developed. To assess the relationship between serum levels of cytokines and concentrations of PM2.5 and PM10 fractions, multiple linear regression was used, gender, age, body mass index (BMI) being included as confounders in the regression models. In addition, a data aggregation method based on the principal component analysis was applied.

To study the relationship between ambient air pollution with particulate matter and bronchial asthma in adults (18–65 years old), a retrospective analysis of the incidence of bronchial asthma (ICD-10 codes J45.0, J45.1, J45.8) for 2014–2020 was carried out. BA incidence was determined for the population of Kazan in and for persons living in the areas at up to 1 km from the monitoring points as well. The database of social and hygienic monitoring and the regional medical information system "Electronic Health of the Republic of Tatarstan" were used. The absolute risks of bronchial asthma in adults (18–65 years old), as well as the absolute risks of BA phenotypes, were calculated. Using linear mixed models based on the Poisson or the negative binomial distribution, the dependences of the absolute risks of BA phenotypes on the PM fraction concentrations were studied.

3. Results and discussions

3.1 Case-control and biomarker studies on the effect of fine particles in the ambient air on atopic allergic and eosinophilic non-allergic phenotypes of the T2-endotype of bronchial asthma in adults

The results of the "case-control" study [33] indicate the role of fine particulate matter in the ambient air in the development of bronchial asthma in adults (18–65 years old), and also suggest the involvement of various underlying mechanisms in the formation of the clinical picture of eosinophilic non-allergic and allergic phenotypes of bronchial asthma: in non-allergic asthma —the reaction of the epithelium to the deposition of particles in the respiratory tract, and in allergic asthma—a reaction to the composition of the aerosol. An increased risk of the eosinophilic non-allergic phenotype of bronchial asthma and its more severe course were noted at higher average annual concentrations of the PM2.5 fraction averaged over 2014–2020. The concentration of bacterial endotoxin had a statistically significant effect on the odds of developing an BA allergic phenotype and was associated with a more severe course of the disease; the odds of an allergic phenotype also increased with an increase of carbon in the composition of the aerosol.

The medians of the average and maximal annual concentrations of PM2.5 fraction averaged over the period 2014–2020 in the areas of residence of patients with bronchial asthma exceeded the maximum allowable levels applied in the Russian Federation (25 and 160 $\mu g/m^3$) by 1.2 and 1.1 times, respectively. For the PM10 fraction, the average annual concentration exceeded the maximum allowable level (40 $\mu g/m^3$) by 2.3 times, and the maximal annual concentration exceeded the maximum allowable level (300 $\mu g/m^3$) by 1.2 times. In the comparison group, the levels of fine PM fractions did not exceed exposure limits, except for the maximum annual concentrations. The chemical composition of the fine PM fractions was represented mainly by carbon (from 36.9–100%) with minor metallic impurities. Contamination with bacterial lipopolysaccharides ranged from 0.0139 to 0.0694 EU/m^3. Microbiological examination of ambient air samples (in the areas of residence of 45 patients with bronchial asthma and 45 persons from the comparison group) showed the growth of bacteria and fungi. Differences in pollution levels between both groups of patients and the comparison group were statistically significant, indicating a higher level of pollution by fine particles, as well as a higher content of carbon and bacterial endotoxin in the areas of residence of patients with bronchial asthma; no differences were found in the total number of microbes.

The study showed the important role of the PM2.5 fraction for patients with eosinophilic non-allergic phenotype of bronchial asthma (**Table 1**): the risk of this bronchial phenotype in adults statistically significantly increases with an increment of the average annual concentration averaged over the period 2014–2020 by 10 $\mu g/m^3$—the odds ratio adjusted by confounders (heredity for asthma, age, concentration of bacterial endotoxin

Exposure parameter	**Adjusted odds ratios (95% confidence intervals)**	
	Non-allergic bronchial asthma	**Allergic bronchial asthma**
PM2.5 average annual concentration averaged over 2014–2020 years. (monitoring data for residential address); increment – 1 $\mu g/m^3$.	4.76 (95% CI: 1.67, 24.40)[1]	OR 4.52 (95% CI 0.91, 55.68)[3]
PM2.5 maximal annual concentration averaged over 2014–2020 yrs. (monitoring data for residential address), increment – 10 $\mu g/m^3$.	1.17 (95% CI: 1.00, 1.42)[2]	1.13 (95% CI: 0.88, 1.58)[3]
PM10 average annual concentration averaged over 2014–2020 years. (monitoring data for residential address), increment – 1 $\mu g/m^3$.	1.71 (95% CI: 1.23, 2.92)[1]	1.84 (95% CI: 0.95, 5.27)[3]
PM10 maximal annual concentration averaged over 2014–2020 years. (monitoring data for residential address), increment – 10 $\mu g/m^3$.	1.12 (95% CI: 1.02, 1.25)[2]	1.11 (95% CI: 0.95, 1.35)[3]
Fraction of carbon in the aerosol composition. Size fraction <3.2 µm (impactor. Measured at the address of residence), increment – 1%.	1.16 (95% CI: 0.98, 1.47)[2]	1.45 (95% CI: 1.02, 2.52)[3]
Bacterial endotoxin in the size fraction <3.2 µm (impactor. Measured at the address of residence), increment – 0.01 EU/m^3.	2.03 (95% CI: 0.81, 6.47)[2]	1.12 (95% CI: 0.94, 1.42)[3]
Bacterial endotoxin in the size fraction 3.2–18 µm (impactor, measured at the address of residence), increment – 0.01 EU/m^3.	3.19 (95% CI: 1.61, 8.51)[2]	1.32 (95% CI: 1.08, 2.00)[3]
Passive smoking, increment – 1 hour/week.	1.90 (95% CI: 0.69, 8.68)[2]	3.24 (95% CI: 1.28, 14.79)[2]

[1]*Confounders: age, heredity for BA, the concentration of bacterial endotoxin in the size fraction <3.2 µm.*
[2]*Confounders: age, heredity for BA.*
[3]*Confounders: heredity for BA, passive smoking.*

Table 1.
Effect of air pollution with particulate matter on the risk of non-allergic and allergic phenotypes of bronchial asthma (case–control study).

in the fraction with a particle size of less than 3.2 μg) was 4.76 (95% CI: 1.67, 24.40); odds ratios characterizing the effect of the PM10 fraction were below 2.0. No statistically significant relationship was found for the allergic phenotype of bronchial asthma and mass concentrations of fine particles in the ambient air. At the same time, the role of bacterial and chemical air pollution in allergic asthma formation was shown: the adjusted odds ratio for an increment of passive smoking duration by 1 hour was 3.24 (95% CI: 1.28, 14.79); for an increment of bacterial endotoxin found in the fraction with deposition in the tracheobronchial region of the respiratory system (3.2–18 μm) by 0.01 EU/m3–1.32 (95% CI: 1.08, 2.00); for an increment of carbon in the chemical composition of the aerosol by 1%—1.45 (95% CI: 1.02, 2.52).

Eosinophilic non-allergic bronchial asthma was better controlled at lower average annual concentrations of the PM2.5 fraction (**Figure 1**), while in the case of allergic asthma, bacterial contamination of the aerosol mattered (**Figure 2**), which may

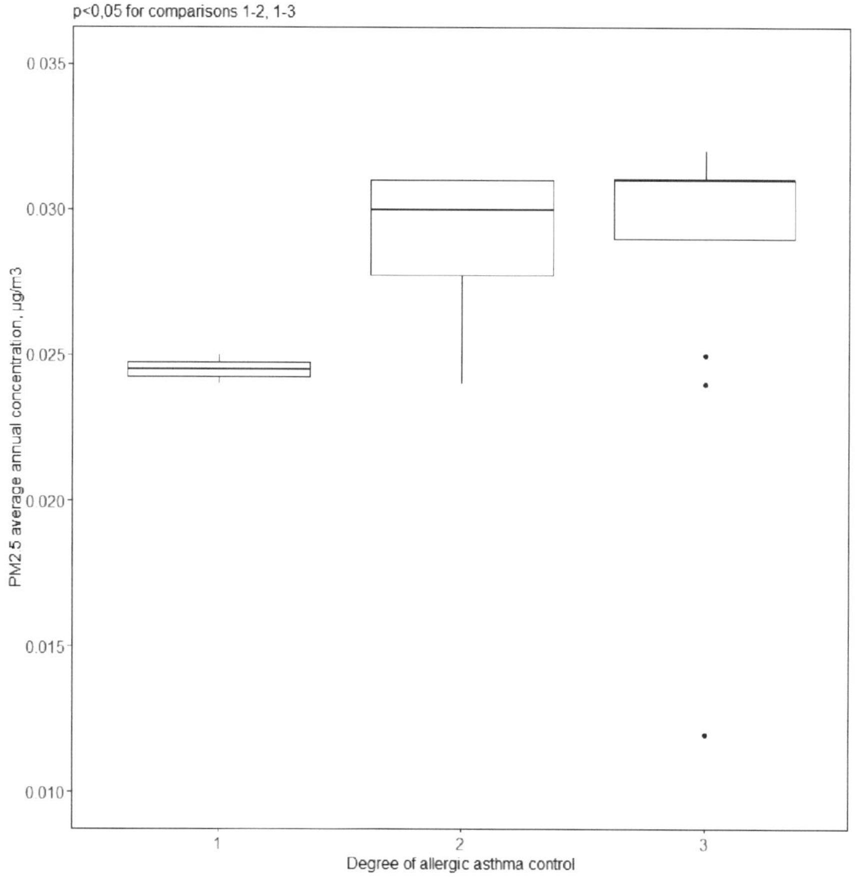

Figure 1.
*PM2.5 average annual concentration (monitoring data for residential zones averaged over the period 2014–2020 in the city of Kazan) for patients with non-allergic and allergic bronchial asthma, depending on the degree of bronchial asthma control. Model 1: PM2.5Avr (mg/m^3) ~ b1i * Degree of control of non-allergic asthma (1 – controlled, 2 – partially controlled, 3 – uncontrolled, Asthma Control Test) + b2 * Age (years) + b3 * Heredity for asthma (no/yes) + b4 * BMI (kg/m^2); b1,1–2 = 0.012, p = 0.09, b1,1–3 = 0.013, p = 0.02. Model 2: PM2.5Avr (mg/m^3) ~ b1i * Degree of control of allergic asthma (1 – controlled, 2 – partially controlled, 3 – uncontrolled) + b2 * Age (number of years) + b3 * Heredity for asthma (no/yes) + b4 * BMI (kg/m^2) + b5 * Passive smoking (hours/week); p > 0.1 for coefficients b1(1–2) and b1(1–3).*

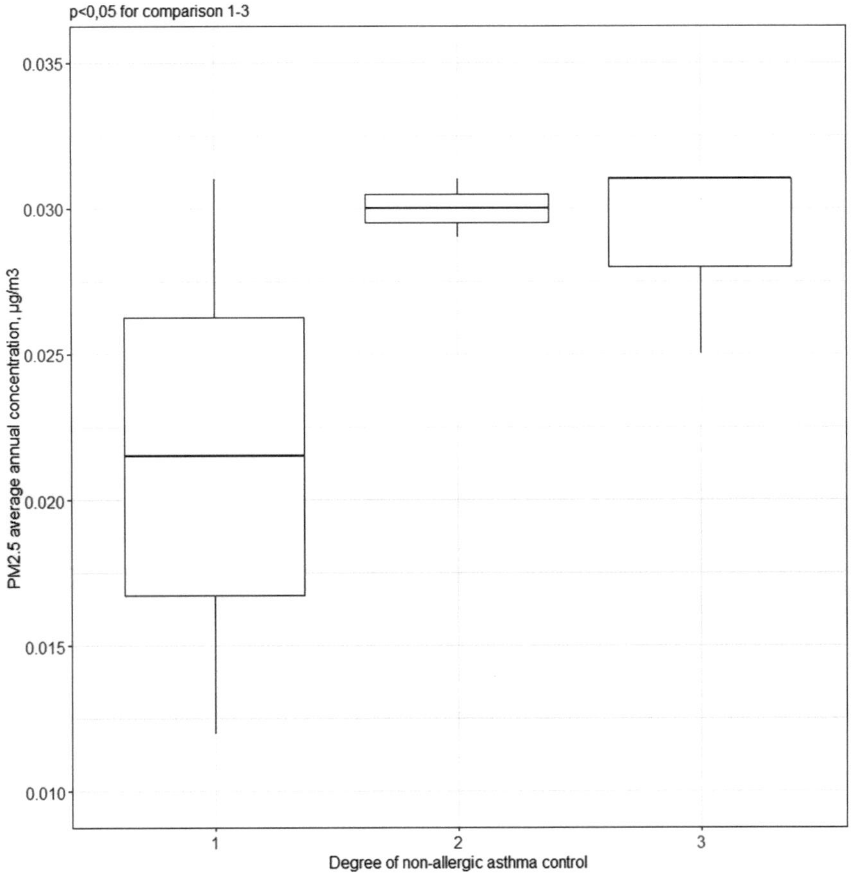

Figure 2.
*Bacterial endotoxin (BE) in the 3.2–18 µm size fraction of ambient particles at the residential zones of patients with non-allergic and allergic bronchial asthma, depending on the degree of disease control. Model 1: BE (EU/m³) ~ b1i * degree of control of non-allergic asthma (1 – controlled, 2 – partially controlled, 3 – uncontrolled) + b2 * age (years) + b3 * heredity for asthma (no/yes) + b4 * BMI (kg/m²); p > 0.1 for coefficients b1,1–2 and b1,1–3. Model 2: BE (EU/m³) ~ b1i * degree of control of allergic asthma (1 – controlled, 2 – partially controlled, 3 – uncontrolled) + b2 * age (years) + b3 * heredity for asthma (no/yes) + b4 * BMI (kg/m²); b1(1–2) = 0.020, p = 0.04, b1(1–3) = 0.027, p = 0.01.*

indicate the importance of various physicochemical characteristics of the suspended solids aerosol in pathogenesis and influence on the clinical course of different phenotypes of the T2-endotype of bronchial asthma.

The data obtained indicate the presence of eosinophilic inflammation in patients of both groups (allergic and eosinophilic non-allergic bronchial asthma) [34]. For patients with eosinophilic non-allergic asthma, an increase in the production of epithelial cytokines IL-33 and IL-25 (alarmins), as well as IL-13 and DPP4, being depended on average annual concentrations of PM2.5 and PM10 averaged over the period 2014–2020. For patients with allergic asthma, similar dependencies were not found. Despite the presence of common signs of T2-type eosinophilic inflammation, the immune patterns of atopic allergic and eosinophilic non-allergic phenotypes of bronchial asthma differed significantly: with a comparable high level of the absolute number of eosinophils in the blood, patients with an allergic phenotype showed pronounced features of the T2-endotype, while with a non-allergic phenotype, there was a lower intensity of T2-type inflammation (IL-4) and an increased expression of genes of cytokines associated with a T17-type response (IL-6, TGF-beta1), being related to the mass of the deposited aerosol. These findings are important for the choice of therapy in different phenotypes of bronchial asthma.

3.2 Ambient particulate matter and bronchial asthma: Results of the epidemiological study based on a geospatial approach

In the epidemiological study based on a geospatial approach [35], a statistically significant increase in the incidence of bronchial asthma was revealed with an increase of 0.09 per 100 annually (growth rate of 17.6% per year). The increase in the incidence of bronchial asthma was observed mainly due to the non-allergic phenotype—by 0.011 per 100 population annually, and the mixed phenotype—by 0.034 per 100 population annually; the increase in the incidence of allergic asthma (by 0.028 per 100 population annually) did not reach statistical significance. An increase in the maximal annual concentrations of PM2.5 by 10 $\mu g/m^3$ increased the absolute risk of non-allergic bronchial asthma by 0.066 per 100 people aged 18–65 years ($p < 0.05$). For other phenotypes of bronchial asthma (allergic asthma, mixed asthma), no statistically significant relationships with mass concentrations and deposited doses were found.

4. Conclusions

As a result of the described studies, the role of ambient fine particles in the development of the eosinophilic non-allergic phenotype of bronchial asthma in adults (18–65 years old) has been proven. The hypothesis about different underlying mechanisms in response to exposure to particulate matter for various phenotypes of bronchial asthma was confirmed. Differences in the immune patterns of the allergic and eosinophilic non-allergic phenotypes of bronchial asthma were established. With a comparable high level of the absolute number of eosinophils in the blood, pronounced features of the T2-endotype were observed in patients with an allergic phenotype. With a non-allergic phenotype, there was a lesser intensity of T2-type inflammation, as well as increased expression of genes of T17-type cytokines, being related to the mass of the deposited aerosol. It has been shown that the estimation of bacterial endotoxin concentrations could be recommended as the most preferable method to characterize the microbiological contamination of atmospheric aerosol in epidemiological studies. Approaches to decision-making in the development of population programs for the prevention of bronchial asthma and in the selection of personalized recommendations for patients with different phenotypes of bronchial asthma were determined. Currently, the treatment of bronchial asthma is based on the use of drugs that can suppress the activity of certain cytokines. The data obtained in the study might become a starting point for the development of new personalized approaches to the treatment and secondary prevention of bronchial asthma associated with air pollution by fine aerosols.

Acknowledgements

The study was funded by RFBR, project number 19-05-50094.

Conflict of interest

The authors declare no conflict of interest.

Author details

Liliya M. Fatkhutdinova[1*], Olesya V. Skorohodkina[1], Laila I. Yapparova[1], Guzel A. Timerbulatova[1] and Ramil R. Zalyalov[2]

1 Kazan State Medical University, Kazan, Russia

2 Republican Medical Center for Information and Analysis, Kazan State Medical University, Kazan, Russia

*Address all correspondence to: liliya.fatkhutdinova@kazangmu.ru

IntechOpen

© 2023 The Author(s). Licensee IntechOpen. This chapter is distributed under the terms of the Creative Commons Attribution License (http://creativecommons.org/licenses/by/3.0), which permits unrestricted use, distribution, and reproduction in any medium, provided the original work is properly cited.

References

[1] Staffoglia M, Oftedal B, Chen J, Rodopoulou S, Rensi M, Atkinson RW, et al. Long-term exposure to low ambient air pollution concentrations and mortality among 28 million people: Results from seven large European cohorts within the ELAPSE project. The Lancet Planetary Health. 2022;**6**(1):e9-e18. DOI: 10.1016/S2542-5196(21)00277-1

[2] Guarnieri M, Balmes JR. Outdoor air pollution and asthma. Lancet. 2014;**383**(9928):1581-1592. DOI: 10.1016/S0140-6736(14)60617-6

[3] Bontinck A, Maes T, Joos G. Asthma and air pollution: Recent insights in pathogenesis and clinical implications. Current Opinion in Pulmonary Medicine. 2020;**26**(1):10-19. DOI: 10.1097/mcp.0000000000000644

[4] Papi A, Brightling C, Pedersen SE, Reddel HK. Asthma. Lancet. 2018;**391**(10122):783-800. DOI: 10.1016/S0140-6736(17)33311-1

[5] To T, Stanojevic S, Moores G, Gershon AS, Bateman ED, Cruz AA, et al. Global asthma prevalence in adults: Findings from the cross-sectional world health survey. BMC Public Health. 2012;**12**:204. DOI: 10.1186/1471-2458-12-204

[6] 2023 GINA Report. Global strategy for asthma management and prevention. Available from: https://ginasthma.org/reports/ [Accessed: May 29, 2023]

[7] Fahy JV. Type 2 inflammation in asthma – Present in most, absent in many. Nature Reviews. Immunology. 2015;**15**(1):57-65. DOI: 10.1038/nri3786

[8] Anderson HR, Favarato G, Atkinson RW. Longterm exposure to air pollution and the incidence of asthma: Meta-analysis of cohort studies. Air Quality, Atmosphere and Health. 2013;**6**:47-56. DOI: 10.1007/s11869-011-0144-5

[9] Bowatte G, Lodge C, Lowe AJ, Erbas B, Perret J, Abramson MJ, et al. The influence of childhood traffic-related air pollution exposure on asthma, allergy and sensitization: A systematic review and a meta-analysis of birth cohort studies. Allergy. 2015;**70**(3):245-256. DOI: 10.1111/all.12561

[10] Khreis H, Kelly C, Tate J, Parslow R, Lucas K, Nieuwenhuijsen M. Exposure to traffic-related air pollution and risk of development of childhood asthma: A systematic review and meta-analysis. Environment International. 2017;**100**:1-31. DOI: 10.1016/j.envint.2016.11.012

[11] Han K, Ran Z, Wang X, Wu Q, Zhan N, Yi Z, et al. Traffic-related organic and inorganic air pollution and risk of development of childhood asthma: A meta-analysis. Environmental Research. 2021;**194**:110493. DOI: 10.1016/j.envres. 2020.110493

[12] McDonnell WF, Abbey DE, Nishino N, Lebowitz MD. Long-term ambient ozone concentration and the incidence of asthma in nonsmoking adults: The AHSMOG study. Environmental Research. 1999;**80**(2 Pt 1):110-121. DOI: 10.1006/enrs.1998.3894

[13] Künzli N, Bridevaux PO, Liu LJ, Garcia-Esteban R, Schindler C, Gerbase MW, et al. Swiss cohort study on air pollution and lung diseases in adults. Traffic-related air pollution correlates with adult-onset asthma among

never-smokers. Thorax. 2009;**64**(8):664-670. DOI: 10.1136/thx.2008.110031

[14] Jacquemin B, Siroux V, Sanchez M, Carsin AE, Schikowski T, Adam M, et al. Ambient air pollution and adult asthma incidence in six European cohorts (ESCAPE). Environmental Health Perspectives. 2015;**123**(6):613-621. DOI: 10.1289/ehp.1408206

[15] Young MT, Sandler DP, DeRoo LA, Vedal S, Kaufman JD, London SJ. Ambient air pollution exposure and incident adult asthma in a nationwide cohort of U.S. women. American Journal of Respiratory and Critical Care Medicine. 2014;**190**(8):914-921. DOI: 10.1164/rccm.201403-0525OC

[16] Requia WJ, Adams MD, Koutrakis P. Association of PM2.5 with diabetes, asthma, and high blood pressure incidence in Canada: A spatiotemporal analysis of the impacts of the energy generation and fuel sales. Science Total Environment. 2017;**584-585**:1077-1083. DOI: 10.1016/j.scitotenv.2017.01.166

[17] Lee DW, Han CW, Hong YC, Oh JM, Bae HJ, Kim S, et al. Long-term exposure to fine particulate matter and incident asthma among elderly adults. Chemosphere. 2021;**272**:129619. DOI: 10.1016/j.chemosphere.2021.129619

[18] Trivedi M, Denton E. Asthma in children and adults-what are the differences and what can they tell us about asthma? Frontiers in Pediatrics. 2019;7:256. DOI: 10.3389/fped.2019.00256

[19] Anenberg SC, Henze DK, Tinney V, Kinney PL, Raich W, Fann N, et al. Estimates of the global burden of ambient PM2.5, ozone, and NO2 on asthma incidence and emergency room visits. Environmental Health Perspectives. 2018;**126**:107004. DOI: 10.1289/EHP3766

[20] Brunekreef B, Strak M, Chen J, Andersen ZJ, Atkinson R, Bauwelinck M, et al. Mortality and morbidity effects of long-term exposure to low-level PM2.5, BC, NO_2, and O_3: An analysis of European cohorts in the ELAPSE project. Research Report in Health Effect Institute. 2021;**2021**(208):1-127

[21] Ścibor M, Malinowska-Cieślik M. The association of exposure to PM10 with the quality of life in adult asthma patients. International Journal of Occupational Medicine and Environmental Health. 2020;**33**(3):311-324. DOI: 10.13075/ijomeh.1896.01527

[22] Fatkhutdinova LM, Tafeeva EA, Timerbulatova GA, Zalyalov RR. Health risks of air pollution with fine particulate matter [in Russian]. Kazan Medical Journal. 2021;**102**(6):862-876. DOI: 10.17816/KMJ2021862

[23] Ai S, Qian ZM, Guo Y, Yang Y, Rolling CA, Liu E, et al. Long-term exposure to ambient fine particles associated with asthma: A cross-sectional study among older adults in six low- and middle-income countries. Environmental Research. 2019;**168**:141-145. DOI: 10.1016/j.envres.2018.09.028

[24] Haldar P, Pavord ID, Shaw DE, Berry MA, Thomas M, Brightling CE, et al. Cluster analysis and clinical asthma phenotypes. American Journal of Respiratory and Critical Care Medicine. 2008;**178**:218-224. DOI: 10.1164/rccm.200711-1754OC

[25] Moore WC, Meyers DA, Wenzel SE, Teague WG, Li H, Li X, et al. Identification of asthma phenotypes using cluster analysis in the severe asthma research program. American Journal of Respiratory and Critical

Care Medicine. 2010;**181**(4):315-323. DOI: 10.1164/rccm.200906-0896OC

[26] Mortaz E, Masjedi MR, Allameh A, Adcock IM. Inflammasome signaling in pathogenesis of lung diseases. Current Pharmaceutical Design. 2012;**18**(16):2320-2328. DOI: 10.2174/138161212800166077

[27] Takizawa H. Impacts of particulate air pollution on asthma: Current understanding and future perspectives. Recent Patents on Inflammation & Allergy Drug Discovery. 2015;**9**(2):128-135. DOI: 10.2174/1872213x09666150623110714

[28] Brusselle G, Bracke K. Targeting immune pathways for therapy in asthma and chronic obstructive pulmonary disease. Annals of the American Thoracic Society. 2014;**11**(Suppl. 5):S322-S328. DOI: 10.1513/AnnalsATS.201403-118AW

[29] Crobeddu B, Aragao-Santiago L, Bui LC, Boland S, Baeza SA. Oxidative potential of particulate matter 2.5 as predictive indicator of cellular stress. Environmental Pollution. 2017;**230**:125-133. DOI: 10.1016/j.envpol.2017.06.051

[30] Mendy A, Wilkerson J, Salo PM, Weir CH, Feinstein L, Zeldin DC, et al. Synergistic association of house endotoxin exposure and ambient air pollution with asthma outcomes. American Journal of Respiratory and Critical Care Medicine. 2019;**200**(6):712-720. DOI: 10.1164/rccm.201809-1733OC

[31] Khan MS, Coulibaly S, Matsumoto T, Yano Y, Miura M, Nagasaka Y, et al. Association of airborne particles, protein, and endotoxin with emergency department visits for asthma in Kyoto, Japan. Environmental Health and Preventive Medicine. 2018;**23**(1):41. DOI: 10.1186/s12199-018-0731-2

[32] De Rooij MMT, Smit LAM, Erbrink HJ, Hagenaars TJ, Hoek G, Ogink NWM, et al. Endotoxin and particulate matter emitted by livestock farms and respiratory health effects in neighboring residents. Environment International. 2019;**132**:105009. DOI: 10.1016/j.envint.2019.105009

[33] Fatkhutdinova LM, Skorohodkina OV, Yapparova LI, KhakimovaM,RakhimzyanovA,AblyaevaA, et al. The effect of fine suspended particles in the atmospheric air on the formation and course of the T2 endotype of bronchial asthma: A case-control [in Russian]. Gigiena i Sanitariya. 2022;**101**(12):1469-1475. DOI: 10.47470/0016-9900-2022-101-12-1469-1475

[34] Skorokhodkina OV, Khakimova MR, Timerbulatova GA, Bareicheva OA, Saleeva LE, Sharipova RG, et al. The role of fine suspended particles of atmospheric air in the formation of eosinophilic inflammation in T2-endotype of asthma [in Russian]. Russian Journal of Allergy. 2022;**19**(4):447-459. DOI: 10.36691/RJA1579

[35] Fatkhutdinova LM, Timerbulatova GA, Zaripov SK, Yapparova L, Ablyaeva A, Saveliev A, et al. Particulate matter in the ambient air as a risk factor of bronchial asthma in adults [in Russian]. Ekologiya Cheloveka (Human Ecology). 2022;**29**(12):875-887. DOI: 10.17816/humeco109943

Chapter 5

A Novel Development in Three-Dimensional Analytical Solutions for Air Pollution Dispersion Modeling

Mehdi Farhane and Otmane Souhar

Abstract

A novel analytical solution has been developed for the three-dimensional dispersion of atmospheric pollutants, providing a new approach to understanding and addressing this important environmental issue. The central concept of the study is to divide the planetary boundary layer into multiple vertical sub-layers, each characterized by its own average wind speed and eddy diffusivity. This allows for a more comprehensive and nuanced examination of atmospheric processes within the boundary layer. The validity of the model is thoroughly evaluated through a comparison of its predictions with data collected from the Copenhagen Diffusion and Prairie Grass experiments. This approach ensures that the model accurately reflects the complexities of atmospheric dispersion in real-world scenarios. The results of the study demonstrate a strong correlation between the predicted and measured crosswind-integrated concentrations. Furthermore, the statistical indices computed for the model fall within an acceptable range, indicating a high level of accuracy in the model's predictions. These findings reinforce the validity of the analytical solution for modeling atmospheric pollutant dispersion.

Keywords: analytical solution, atmospheric dispersion, atmospheric boundary layer sub-layers, Fourier transform, Sturm-Liouville eigenvalue problem

1. Introduction

The scientific community has been driven to act urgently due to the increase in ecological disasters worldwide. A key approach to addressing this issue is the development of reliable models that enable quantitative prediction of pollution-related phenomena, either through analytical descriptions or simulations using powerful and operational tools. These models, which can be based on simulations or analytical descriptions, have strong quantitative predictive power for understanding pollution-related phenomena. The atmosphere is considered the primary means of dispersing pollutants in the environment, which can come from

IntechOpen

industrial sources or accidental events, leading to progressive contamination of the ecosystem, fauna, flora, and all populations. Therefore, accurately evaluating how pollutants move through the atmosphere in the boundary layer is vital to preserve, protect, and restore the integrity of ecosystems. To accomplish this, it is essential to create atmospheric dispersion models tailored to the parameters and weather conditions specific to the region being studied, using parameters, weather conditions, and local topographic information. These models should produce realistic outcomes on environmental consequences, helping to minimize the negative impact of potential disasters like forest fires. Furthermore, implementing simulations of models based on specific cases can be beneficial in establishing particular emission limits for industrial sites, limiting the release of pollutants into the air.

The aim of our work is to propose a closed-form analytical solution for three-dimensional advection-dispersion transport problems in finite, multilayered media using rigorous mathematical tools. We ensure that wind velocity profiles and vertical diffusivity coefficients take average values in each sub-layer. We have made innovative contributions by using a governing function to generate the eigenvalues associated with our problem and overcoming the common difficulty of missing some of the eigenvalues when they are calculated by developing a transcendental equation for each layer. Finally, our approach has enabled us to obtain a closed-form analytical solution that differs from previous solutions that required the determination of integral coefficients. This method could have important implications in many scientific and technical fields for solving complex advection-dispersion transport problems in finite, multilayered media.

In recent years, there has been an increasing emphasis on creating new analytical approaches that can be used for a range of wind speeds and turbulent diffusivity coefficients. For the most part, these approaches involve the projection of the solution onto a basis of orthogonal polynomials, such as the GILTT technique [1–8]. However, an important drawback of this method is that it requires a large number of eigenvalues to ensure convergence, which can reach up to 250 eigenvalues. In contrast, our new solution is able to provide better results by using a much smaller number of eigenvalues, specifically between 10 and 15, to ensure convergence. This is a significant improvement over existing approaches, making our method more efficient and practical for analytical applications.

Our model was developed using the Fourier transform and separation of variables technique, which resulted in the Sturm-Liouville problem. In order to understand the factors influencing pollutant dispersion, we considered the following parameters: (*i*) the wind speed profile of Deaves and Harris [9], (*ii*) the vertical turbulent diffusivity coefficient, which is considered as an explicit function of the downwind distance and vertical height under convective conditions, as described by Mooney and Wilson [10] and Degrazia et al. [11], and (*iii*) the lateral eddy diffusivity coefficient, which also depends on the downwind distance and vertical height, as described in the work of Huang [12] and Brown et al. [13]. We first proceed to the exposition of the general formulation of the equation that governs the dispersion of pollutants in the atmospheric boundary layer, with a comprehensive presentation in Section 2. We then turn to the explicit solution in Section 3, while exploring the model parameterizations in Section 4. We provide an in-depth discussion of our numerical results in Section 5. Finally, the last section proposes our conclusion, a synthesis of our investigation.

2. Presentation of the general problem

The movement of pollutants within the planetary boundary layer (PBL) is characterized by turbulent dispersion, which is governed by a mathematical equation known as the advection-diffusion equation. This equation considers the combined influence of two primary mechanisms, advection and diffusion, which respectively refer to the transport of pollutants by wind and the spreading of pollutants due to turbulence.

By applying the advection-diffusion equation, researchers can gain a more comprehensive understanding of the intricate and dynamic behavior of pollutants within the PBL. This equation serves as a mathematical framework that accounts for various factors that affect pollutant dispersion, such as wind direction and speed, Eddy diffusivities, and atmospheric stability, and pollutant concentration, presented in the following form:

$$\frac{\partial C}{\partial t} + \nabla.(\mathbf{U_w}\ C) = \nabla.(D\ \ \nabla C) + S, \tag{1}$$

where $\mathbf{U_w} = (U, V, W)^T$ is the wind speed vector (m/s) representing the components U, V, and W in the east-west, north-south and vertical directions, respectively; D is the molecular diffusion coefficient; S is the source term; and ∇ is the gradient operator.

By use of the time average and fluctuation values, $U = u + u'$, $V = v + v'$, $W = w + w'$ and $C = c + c'$, the wind speed vector $\mathbf{U_w}$ is expressed as:

$$\mathbf{U_w} = \overline{\mathbf{U_w}} + \mathbf{U_w}', \quad \text{with} \quad \overline{\mathbf{U_w}} = (u, v, w)^T \quad \text{and} \quad \mathbf{U_w}' = (u', v', w')^T \tag{2}$$

Applying the Reynolds averaging rules [14] to the vertical mass flow term, denoted as $\mathbf{U_w}\ C$, results in the following expression: It turns out that turbulent diffusion can be described with Fick's laws of diffusion as follows [15].

$$\overline{u'c'} = -K_x\ \frac{\partial c}{\partial x}; \quad \overline{v'c'} = -K_y\ \frac{\partial c}{\partial y}; \quad \overline{w'c'} = -K_z\ \frac{\partial c}{\partial z}. \tag{3}$$

where K_x, K_y, and K_z are the eddy diffusivities components along x, y, and z directions, respectively.

It should be noted that in a turbulent boundary layer where advection is occurring, K will be larger than D and eddy diffusion will dominate solute transport. In this case, the molecular diffusion coefficient $\nabla.(D\ \ \nabla C)$ is then to be replaced by an eddy or turbulent diffusivity. The source term could be eliminated from Eq. (1) and should be added to the boundary conditions as a delta function: At the point $(0, 0, H_s)$, there is a source rejecting the pollutant with a continuous flow Q:

$$u\ \ c(0, y, z) = Q\ \ \delta(y) \otimes \delta(z - H_s), \tag{4}$$

where $\otimes$ is the tensor product of two distributions and H_s is the source height. By application of the Reynolds averaging and the divergence operator to Eq. (3), Eq. (1) may be written as follows.

$$\frac{\partial c}{\partial t} = -\frac{\partial}{\partial x}\overline{u'c'} - \frac{\partial}{\partial y}\overline{v'c'} - \frac{\partial}{\partial z}\overline{w'c'} - \overline{U_w}.\nabla c. \tag{5}$$

In the remainder of this paper, the following assumptions are considered:

a. the steady state condition (i.e., $\frac{\partial c}{\partial t} = 0$);

b. the two terms $v\frac{\partial c}{\partial y}$ and $w\frac{\partial c}{\partial z}$ are neglected since the x-axis coincides with the wind flow average, therefore the w and v wind velocity components are less important; and

c. the turbulent diffusion in the direction of the mean wind is neglected compared to the advection transport mechanism (i.e., $u\frac{\partial c}{\partial x} >> \frac{\partial}{\partial x}\left(K_x\frac{\partial c}{\partial x}\right)$).

These assumptions lead to the steady-state advection-diffusion equation defined as:

$$\begin{cases} u(z)\dfrac{\partial c}{\partial x} = \dfrac{\partial}{\partial y}\left(K_y\dfrac{\partial c}{\partial y}\right) + \dfrac{\partial}{\partial z}\left(K_z\dfrac{\partial c}{\partial z}\right); & (x,y,z)\in]0,L_x[\times]-L_y,L_y[\times]0,H_{mix}[, \\ \text{subject to the boundary conditions :} \begin{cases} u(z)\ c(x,y,z) = Q\cdot\delta(y)\otimes\delta(z-H_s); & x=0\ , \\ K_y(x,z)\dfrac{\partial c(x,y,z)}{\partial y} \to 0; & |y|\to L_y, \\ K_z(x,z)\dfrac{\partial c(x,y,z)}{\partial z} = 0; & z\ \in\{0,H_{mix}\}. \end{cases} \end{cases} \tag{6}$$

where z_0 is the surface roughness length and H_{mix} is the PBL height.
We consider that the eddy diffusivities have the following separable formulations:

$$K_y(x,z) = \zeta_y(x)\ u(z), \tag{7}$$

$$K_z(x,z) = \xi(x)\ \varphi_z(z). \tag{8}$$

We vertically divide the PBL into H intervals such that for each one the eddy diffusivity and wind speed assume average values. For $h = 1,\cdots,H$,

$$u_h = \frac{1}{z_h - z_{h-1}}\int_{z_{h-1}}^{z_h} u(s)ds, \tag{9}$$

$$\varphi_{z_h} = \frac{1}{z_h - z_{h-1}}\int_{z_{h-1}}^{z_h} \varphi_z(s)ds. \tag{10}$$

By use of the formulations of K_y and K_z given by Eqs. (7) and (8), Eq. (5) is written as:

$$u_h\ \frac{\partial c_h}{\partial x} = \zeta_y(x)\ u_h\ \frac{\partial^2 c_h}{\partial y^2} + \xi(x)\ \varphi_{z_h}\ \frac{\partial^2 c_h}{\partial z^2}, \tag{11}$$

with u_h and φ_{z_h} (given by Eqs. (9) and (10)) are constants.
This later Eq. (11) is subject to the first boundary conditions Eq. (6) on the one hand, and on the other hand, the continuity of both the concentration and the flux at the interface level is applied.

$$\begin{cases} \varphi_{z_1} \dfrac{\partial c_1(x,y,z_0)}{\partial z} = 0, \\ \begin{cases} c_{h-1}(x,y,z_{h-1}) = c_h(x,y,z_{h-1}) \\ \varphi_{z_{h-1}} \dfrac{\partial c_{h-1}(x,y,z_{h-1})}{\partial z} = \varphi_{z_h} \dfrac{\partial c_h(x,y,z_{h-1})}{\partial z} \end{cases}, \quad \forall h \in \{2,\cdots,H\}, \\ \varphi_{z_H} \dfrac{\partial c_H(x,y,z_H)}{\partial z} = 0 \end{cases}$$

3. A mathematical approach to solving the pollutant dispersion equation: analytical solution

We start this section by applying Fourier transform to Eq. (11). Let $\hat{c}_h^{\omega}(x,z)$ denote the Fourier transformation of c_h with respect to y.

$$\hat{c}_h^{\omega}(x,z) = \int_{-\infty}^{+\infty} c_h(x,y,z) e^{-2i\pi\omega y} dy, \quad h \in \{1,\cdots,H\}$$

which gives

$$u_h \left[\frac{\partial \hat{c}_h^{\omega}}{\partial x} + (2\pi)^2 \omega^2 \zeta_y(x) \hat{c}_h^{\omega} \right] = \xi(x) \ \varphi_{z_h} \frac{\partial^2 \hat{c}_h^{\omega}}{\partial z^2}, \quad z \in [z_{h-1}, z_h]. \tag{12}$$

Let

$$\chi_h^{\omega}(x,z) = \hat{c}_h^{\omega}(x,z).\exp\left((2\pi)^2 \ \omega^2 \int_0^x \zeta_y(s) ds \right) \tag{13}$$

then

$$\frac{\partial \chi_h^{\omega}}{\partial x} = \left[\frac{\partial \hat{c}_h^{\omega}}{\partial x} + (2\pi)^2 \omega^2 \zeta_y(x) \hat{c}_h^{\omega} \right] \ \exp\left((2\pi)^2 \ \omega^2 \int_0^x \zeta_y(s) ds \right).$$

By multiplying both sides of Eq. (12) by $\exp\left((2\pi)^2 \ \omega^2 \int_0^x \zeta_y(s) ds \right)$ and since u_h and φ_{z_h} are constants for each interval, we show easily that, for all $h \in \{1,\cdots,H\}$:

$$u_h \frac{\partial \chi_h^{\omega}}{\partial x} = \xi(x) \ \varphi_{z_h} \frac{\partial^2 \chi_h^{\omega}}{\partial z^2}, \quad z \in [z_{h-1}, z_h] \tag{14}$$

We proceed in the same way with the boundary conditions, and find

$$\begin{cases} \varphi_{z_1} \dfrac{\partial \chi_1^{\omega}(x,z_0)}{\partial z} = 0 \\ \begin{cases} \chi_{h-1}^{\omega}(x,z_{h-1}) = \chi_h^{\omega}(x,z_{h-1}) \\ \varphi_{z_{h-1}} \dfrac{\partial \chi_{h-1}^{\omega}(x,z_{h-1})}{\partial z} = \varphi_{z_h} \dfrac{\partial \chi_h^{\omega}(x,z_{h-1})}{\partial z} \end{cases}, h \in \{2,\cdots,H\} \\ \varphi_{z_H} \dfrac{\partial \chi_H^{\omega}(x,z_H)}{\partial z} = 0 \end{cases}$$

The solution of Eq. (14) is assumed to be in the form:

$$\chi_h^{\omega}(x,z) = \sum_{n=0}^{\infty} G_{h,n}^{\omega}(x)\ P_{h,n}(z), \qquad h \in \{1, \cdots, H\}$$

This separated form gives two ordinary differential equations to be solved:

$$\frac{d\ G_{h,n}^{\omega}}{dx} + \gamma_n^2\ \xi(x)\ G_{h,n}^{\omega} = 0 \tag{15}$$

and

$$\varphi_{z_h}\ \frac{d^2\ P_{h,n}}{dz^2} + \gamma_n^2\ u_h\ P_{h,n} = 0, \tag{16}$$

where γ_n is a separation constant.

The first-order ordinary differential Eq. (15) has the solution

$$G_{h,n}^{\omega}(x) = \mu_n(\omega)\ .\exp\left(-\gamma_n^2\ \int_0^x \xi(s)ds\right),$$

where μ_n is an arbitrary function depending on ω.

Eq. (16) represents a Sturm-Liouville problem. Solutions of such problem form an eigenfunction basis of the form:

$$P_{h,n}(z) = \alpha_{h,n}\cos(\lambda_{h,n}z) + \beta_{h,n}\sin(\lambda_{h,n}z), \tag{17}$$

where, $\lambda_{h,n} = \gamma_n\ \sqrt{\frac{u_h}{\varphi_{z_h}}}$.

The Eq. (17) satisfies the following boundary conditions:

$$\begin{cases} \varphi_{z_1}\dfrac{dP_{1,n}(z_0)}{dz} = 0 \quad (a) \\ \begin{cases} P_{h-1,n}(z_{h-1}) = P_{h,n}(z_{h-1}) \\ \varphi_{z_{h-1}}\dfrac{dP_{h-1,n}(z_{h-1})}{dz} = \varphi_{z_h}\dfrac{dP_{h,n}(z_{h-1})}{dz} \end{cases}, h \in \{2, \cdots, H\} (b) \\ \varphi_{z_H}\dfrac{dP_{H,n}(z_H)}{dz} = 0 \quad (c) \end{cases} \tag{18}$$

To calculate the expression of $P_{h,n}$, it comes down to calculate the values of $\alpha_{h,n}$ and $\beta_{h,n}$, on each of the sub-layer $[z_{h-1}, z_h]$, $h \in \{1, \cdots, H\}$.

By solving the recursive system resulting from substitution of Eq. (17) in Eq. (18), we obtain respectively the formulations of $\alpha_{h,n}$ and $\beta_{h,n}$. More specifically,

For the first sub-layer, $\alpha_{1,n}$ and $\beta_{1,n}$ satisfy the equation:

$$\alpha_{1,n}\sin(\lambda_{1,n}z_0) - \beta_{1,n}\cos(\lambda_{1,n}z_0) = 0 \tag{19}$$

from which, we can take $\beta_{1,n} = \sin(\lambda_{1,n}z_0)$, so that $\alpha_{1,n} = \cos(\lambda_{1,n}z_0)$, which means:

$$P_{1,n}(z) = \cos(\lambda_{1,n}(z - z_0)).$$

For the last sub-layer (H^{th} sub-layer):

$$\alpha_{H,n} = \cot(\lambda_{H,n} z_H)\ \beta_{H,n}. \tag{20}$$

And for the intermediate sub-layers:

$$\begin{aligned}\alpha_{h,n} = & \frac{\varphi_{z_h}\lambda_{h,n} - \varphi_{z_{h-1}}\lambda_{h-1,n}}{2\varphi_{z_h}\lambda_{h,n}} \left[\left(\cos((\lambda_{h,n} + \lambda_{h-1,n}) z_{h-1}) + \frac{\varphi_{z_h}\lambda_{h,n} + \varphi_{z_{h-1}}\lambda_{h-1,n}}{\varphi_{z_h}\lambda_{h,n} - \varphi_{z_{h-1}}\lambda_{h-1,n}} \cdot \right.\right. \\ & \cos((\lambda_{h,n} - \lambda_{h-1,n}) z_{h-1}))\alpha_{h-1,n} + (\sin((\lambda_{h,n} + \lambda_{h-1,n}) z_{h-1}) - \\ & \left.\left. \frac{\varphi_{z_h}\lambda_{h,n} + \varphi_{z_{h-1}}\lambda_{h-1,n}}{\varphi_{z_h}\lambda_{h,n} - \varphi_{z_{h-1}}\lambda_{h-1,n}} \sin((\lambda_{h,n} - \lambda_{h-1,n}) z_{h-1}) \right) \beta_{h-1,n} \right],\end{aligned} \tag{21}$$

and

$$\begin{aligned}\beta_{h,n} = & \frac{\varphi_{z_h}\lambda_{h,n} - \varphi_{z_{h-1}}\lambda_{h-1,n}}{2\varphi_{z_h}\lambda_{h,n}} \left[\left(\sin((\lambda_{h,n} + \lambda_{h-1,n}) z_{h-1}) + \frac{\varphi_{z_h}\lambda_{h,n} + \varphi_{z_{h-1}}\lambda_{h-1,n}}{\varphi_{z_h}\lambda_{h,n} - \varphi_{z_{h-1}}\lambda_{h-1,n}} \cdot \right.\right. \\ & \sin((\lambda_{h,n} - \lambda_{h-1,n}) z_{h-1}))\alpha_{h-1,n} - (\cos((\lambda_{h,n} + \lambda_{h-1,n}) z_{h-1}) - \\ & \left.\left. \frac{\varphi_{z_h}\lambda_{h,n} + \varphi_{z_{h-1}}\lambda_{h-1,n}}{\varphi_{z_h}\lambda_{h,n} - \varphi_{z_{h-1}}\lambda_{h-1,n}} \cos((\lambda_{h,n} - \lambda_{h-1,n}) z_{h-1}) \right) \beta_{h-1,n} \right].\end{aligned} \tag{22}$$

The eigenvalues $\gamma_n,\ n \in \mathbb{N}^*$ of this problem are real and discrete and the eigenfunctions are mutually orthogonal. The orthogonality relation developed by [16] for this class of (self-adjoint) problems with respect to the density u_h on each interval $[z_{h-1}, z_h],\ h \in \{1, \cdots, H\}$ leads to

$$\sum_{h=1}^{H} \int_{z_{h-1}}^{z_h} u_h P_{h,m}(s) P_{h,n}(s)\ ds = \|P_{h,m}\|.\ \|P_{h,n}\|.\ \delta_{m,n},$$

where $\delta_{m,n}$ is the Kronecker symbol. Then, $\|P_{h,n}\|^2$ is written as follows.

$$\begin{aligned}\|P_{H,n}\|^2 &= \sum_{h=1}^{H} \int_{z_{h-1}}^{z_h} u_h\, (P_{h,n}(s))^2 ds \\ &= \begin{cases} \displaystyle\sum_{h=1}^{H} u_h (z_h - z_{h-1}), \quad \text{when } \gamma_n = 0 \\ \displaystyle\sum_{h=1}^{H} \frac{u_h}{2\ \lambda_{h,n}} [\sin(\lambda_{h,n}(z_h - z_{h-1}))((\alpha_{h,n}^2 - \beta_{h,n}^2)\cos(\lambda_{h,n}(z_h + z_{h-1})) + \\ 2\alpha_{h,n}\beta_{h,n}\sin(\lambda_{h,n}(z_h + z_{h-1}))) + \lambda_{h,n}(\alpha_{h,n}^2 + \beta_{h,n}^2)(z_h - z_{h-1})], \quad \text{when } \gamma_n \neq 0. \end{cases}\end{aligned}$$

The eigenvalues of each sub-layer can be obtained by integrating Eq. (17) on each of the intervals $[z_{h-1}, z_h],\ h \in \{1, \cdots, H\}$ taking into account the boundary conditions Eqs. (19)–(21). Whereas the eigenfunctions $P_{H,n}$ form a complete set, so χ_H^{ω} can be developed as:

$$\chi_H^{\omega}(x, z) = \sum_{n=0}^{\infty} a_n\ \exp\left(-\gamma_n^2 \int_0^x \xi(s) ds\right)\ P_{H,n}(z).$$

Then, the coefficients a_n are given by:

$$
\begin{aligned}
a_n &= \frac{1}{\|P_{h,n}\|^2} \sum_{\ell=1}^{H} \int_{z_{\ell-1}}^{z_\ell} u_\ell \chi_\ell^{\omega}(0,z) P_{\ell,n}(z) dz \\
&= \begin{cases} \dfrac{Q}{\sum_{h=1}^{H} u_h (z_h - z_{h-1})}, & \text{when } \gamma_n = 0 \\ \dfrac{Q}{\|P_{h,n}\|^2} \sum_{\ell=1}^{H} P_{\ell,n}(H_s), & \text{when } \gamma_n \neq 0 \end{cases}
\end{aligned}
$$

The inverse Fourier transform of $\exp\left((2\pi)^2 \ \omega^2 \int_0^x \zeta_y(s) ds\right)$ is:

$$
\frac{1}{2\sqrt{\pi \displaystyle\int_0^x \zeta_y(s) ds}} \exp\left(- \frac{y^2}{4 \displaystyle\int_0^x \zeta_y(s) ds} \right)
$$

Consequently,

$$
c_H(x,y,z) = \frac{Q}{2\sqrt{\pi \displaystyle\int_0^x \zeta_y(s) ds}} \exp\left(- \frac{y^2}{4 \displaystyle\int_0^x \zeta_y(s) ds} \right) \cdot \\
\sum_{n=0}^{\infty} \exp\left(-\gamma_n^2 \int_0^x \xi(s) ds \right) \frac{P_{H,n}(z)}{\|P_{H,n}\|^2} \left(\sum_{h=1}^{H} P_{h,n}(H_s) \right). \tag{23}
$$

The Crosswind-Integrated Concentration (CIC) model is a valuable tool for analyzing the behavior of pollutants in the atmosphere. It takes into account the horizontal wind component, which is a significant factor in pollutant transport. Mathematically, the crosswind-integrated concentration $c_H^y(x,z)$ is obtained by integrating $c_H(x,y,z)$ in Eq. (20) to y from $-\infty$ to $+\infty$:

$$
c_H^y(x,z) = \int_{-\infty}^{+\infty} c_H(x,y,z) dy
$$

The (CIC) model considers the dispersion of pollutants in the crosswind direction due to the horizontal wind component and produces a two-dimensional concentration map that can be used to assess the impact of pollutants in both the downwind (x-axis) and crosswind (z-axis) directions by integrating the concentration of pollutants over the crosswind direction (y), it gives

$$
c_H^y(x,z) = Q \sum_{n=0}^{\infty} \exp\left(-\gamma_n^2 \int_0^x \xi(s) ds \right) \frac{P_{H,n}(z)}{\|P_{H,n}\|^2} \sum_{h=1}^{H} P_{h,n}(H_s). \tag{24}
$$

4. Parameterizing pollutant dispersion: Exploring model inputs

The wind speed profile $u(z)$ is determined using the Deaves and Harris model [9], which is based on experimental measurements of wind speed profiles. This model is advantageous because it provides a more accurate representation of the logarithmic law at low heights, and improves the accuracy of both the logarithmic and power law models at moderate heights. The profile is given by the equation

$$u(z) = \frac{u_*}{k}\left[\ln\left(\frac{z+z_0}{z_0}\right) + 5.75\left(\frac{z}{H_{mix}}\right) - 1.88\left(\frac{z}{H_{mix}}\right)^2 - 1.33\left(\frac{z}{H_{mix}}\right)^3 + 0.25\left(\frac{z}{H_{mix}}\right)^4\right], \tag{25}$$

where u_* is the friction velocity, k is the von Karman constant, z_0 is the roughness length, and H_{mix} is the height of the planetary boundary layer (PBL). The Deaves and Harris model includes terms that account for the effects of turbulence and thermal stratification on the wind profile. By using this model, we can better understand and predict the behavior of the wind at different heights within the PBL.

The modified form of the vertical eddy diffusivity coefficient K_z, given in Eq.(8), is adopted from [11], where the functional $\varphi_z(z)$ is expressed as:

$$\varphi_z(z) = 0.22 H_{mix} w_* \left(\frac{z}{H_{mix}}\left(1-\frac{z}{H_{mix}}\right)\right)^{1/3}\left(1 - \exp\left(-\frac{4z}{H_{mix}}\right) - 0.0003\exp\left(\frac{8z}{H_{mix}}\right)\right), \tag{26}$$

where w_* is the convective velocity. The integrable correction dimensionless function in Eq. (8) is defined in terms of the along-wind length scale L_1 as follows [10]:

$$\xi(x) = 1 - \exp\left(-\frac{x}{L_1}\right). \tag{27}$$

The length L_1 is given in terms of u, φ_z and σ_w as [10]:

$$L_1 = \frac{1}{\sigma_w^2(H_s)} u(H_s)\varphi_z(H_s),$$

where σ_w is the vertical turbulent intensity. Note that, there exist many expressions of σ_w, we adopt here the expression given by [17]:

$$\sigma_w = 0.96 w_* \left(\frac{3z}{H_{mix}} + \frac{|L|}{H_{mix}}\right)^{1/3},$$

The Monin-Obukhov length (L) is a commonly parameter used in atmospheric dispersion modeling [18]. It plays a crucial role in characterizing the stability of the atmospheric boundary layer, which is essential for understanding and predicting the dispersion of pollutants or particles in the atmosphere. Specifically, the sign of L indicates the atmospheric stability state. A negative value of L indicates an unstable atmosphere, where turbulence is generated due to buoyancy forces. A positive value of L indicates a stable atmosphere, where turbulence is suppressed due to stratification. A large absolute value of L (i.e., $|L| \gg 1$) indicates a neutral atmosphere, where turbulence is neither generated nor suppressed due to buoyancy effects [18].

It should be noted that when $L_1 \to 0$, the term $\exp\left(-\frac{x}{L_1}\right)$ becomes negligible in (27), and K_z will dependent only on z (i.e., $K_z \equiv \varphi_z$ in Eq. (8)).

The lateral eddy diffusivity is given by [12]:

$$K_y(x,z) = \frac{1}{2}\ u(z)\frac{d\sigma_y^2(x)}{dx}, \tag{28}$$

where σ_y is the standard deviation in the crosswind direction (depends only on x). By identifying Eqs. (7) and (28), we obtain

$$\zeta_y(x) = \frac{1}{2}\ \frac{d\sigma_y^2(x)}{dx}$$

5. Validating analytical solutions for air pollution dispersion modeling: analyzing experimental data

This section offers a comprehensive analysis of the data generated by our model, which is founded on the Deaves and Harris wind velocity profile. The model is constructed using two different formulations of vertical turbulent diffusivity. The first formulation, with $L_1 > 0$, factors in both the downwind distance and the vertical height, while the second formulation, with $L_1 \to 0$, considers only the vertical height. The performance of the model, as presented in Eq. (24), is assessed and confirmed through the use of datasets from the Copenhagen diffusion experiments in Denmark and the Prairie Grass experiments in the USA.

Atmospheric dispersion experiments were conducted in Copenhagen, Denmark, from June 27 to November 9, 1978, to study air pollutant transport in the lower atmosphere. The tracer gas used in these experiments was sulfur hexafluoride (SF6), which was released from a tower located at a height of 115 meters (H_s), with a roughness length of 0.6 meters (z_0). Additional details of the parameters used in these experiments can be found in the publications of Gryning and Lyck [19] and Gryning et al. [20].

The Prairie Grass experiments, carried out near O'Neil, Nebraska, between March and August 1956, were a groundbreaking series of studies on atmospheric pollution behavior and transport. These experiments measured the dispersion of sulfur dioxide (SO_2) released from a 1.5 meter tower at five downwind distances from the source: 50, 100, 200, 400, and 800 meters. The roughness length of the area was 0.006 meters, which is critical in determining the spread of air pollutants. The experiments were extensively documented by Barad et al. [21] and Nieuwstadt et al. [22], providing essential initial insights into atmospheric dispersion processes.

Overall, the Copenhagen and Prairie Grass experiments provided essential knowledge on the transport and dispersion of air pollutants in the lower atmosphere. The Copenhagen experiments used an inert tracer gas to examine general dispersion patterns, while the Prairie Grass experiments focused on the downwind transport of SO_2, providing dispersion data across a range of distances. Both experiments relied on the roughness length of the terrain, a crucial parameter in determining how pollutants spread near the ground. The findings of these landmark experiments made significant contributions to scientists' understanding of atmospheric pollution behavior, particularly in the PBL near the Earth's surface.

It should be noted that the computations were carried out using two distinct programming environments, namely MATLAB and Wolfram Mathematica. Specifically, the function fzero, which utilizes a combination of the bisection, secant, and inverse quadratic interpolation methods, was defined and executed in the MATLAB environment. Conversely, the FindRoot function, which combines Brent's and Newton's methods along with a method that approximates the Jacobian, was defined and executed in the Wolfram Mathematica environment.

To implement the analytical solution in a practical sense, we have employed a discretization technique for the atmospheric boundary layer. Specifically, we have discretized the layer into two and four sublayers, respectively. In the case of two sub-layers, the boundary layer height is discretized using the following approach: For the two sub-layers case, the boundary layer height is divided into two sub-layers, and the discretization is taken as follows: $dz_1 = \frac{7}{13}\ (H_{mix} - z_0)$ and $dz_2 = \frac{6}{13}\ (H_{mix} - z_0)$. On the other hand, for the four sub-layers case, the boundary layer height is divided into four sub-layers, and the discretization is taken as follows: $dz_1 = \frac{3}{15}\ (H_{mix} - z_0)$, $dz_2 = \frac{3}{15}(H_{mix} - z_0)$, $dz_3 = \frac{7}{15}\ (H_{mix} - z_0)$, and $dz_4 = \frac{2}{15}\ (H_{mix} - z_0)$. These formulas are used to accurately model the behavior of the boundary layer height in numerical simulations.

In this work, the results of the numerical convergence for concentration predictions of atmospheric dispersion models are indicated as a function of the number of eigenvalues used in the model for different downwind distances from the source. Reference is made to experiments conducted in Copenhagen (experiments 1 to 3) and Prairie Grass (experiments 1, 5, and 7).

For the Copenhagen experiments, the models used four sub-layers in the vertical and tested two different formulations for calculating the vertical eddy diffusivity, a parameter that represents turbulence and mixing in the vertical direction. **Figure 1** shows the numerical convergence of normalized crosswind-integrated concentration as a function of the number of eigenvalues used in the model. The convergence is shown for four different downwind distances (200, 500, 1000, and 1500 m) from the source. The sentence indicates that for Prairie Grass, only one formulation of vertical diffusivity was used because the mixing height (H_mix) was much greater than the surface layer height (H_s), so the two formulations gave the same result.

Figure 2 shows the numerical convergence of ground-level crosswind-integrated concentration for Prairie Grass experiments as a function of the number of eigenvalues at four downwind distances (50, 100, 200, and 800 m).

Numerical simulations were conducted to investigate the influence of the vertical eddy diffusivity formulation on the normalized crosswind-integrated concentration of air pollution near a point source. Specifically, the simulations were performed for Copenhagen experiments numbered 1, 2, and 3, and the results are presented in **Figure 3**.

The figure depicts the normalized crosswind-integrated concentration $\frac{c_H^y(x, z_0)}{Q}$ at ground level as a function of downwind distance, using two different vertical turbulence formulations. The results indicate that the choice of vertical eddy diffusivity formulation has a significant impact on the concentration near the point source. Notably, the figure shows that the concentration is higher when using the first formulation of the vertical eddy diffusivity, compared to the second one. Furthermore, the figure illustrates the expected decrease in crosswind-integrated concentration with increasing downwind distance due to the dispersion of air pollutants caused by turbulent mixing in the atmosphere. In conclusion, these simulations provide valuable insights into air pollution dispersion behavior and can inform future experiments and modeling efforts.

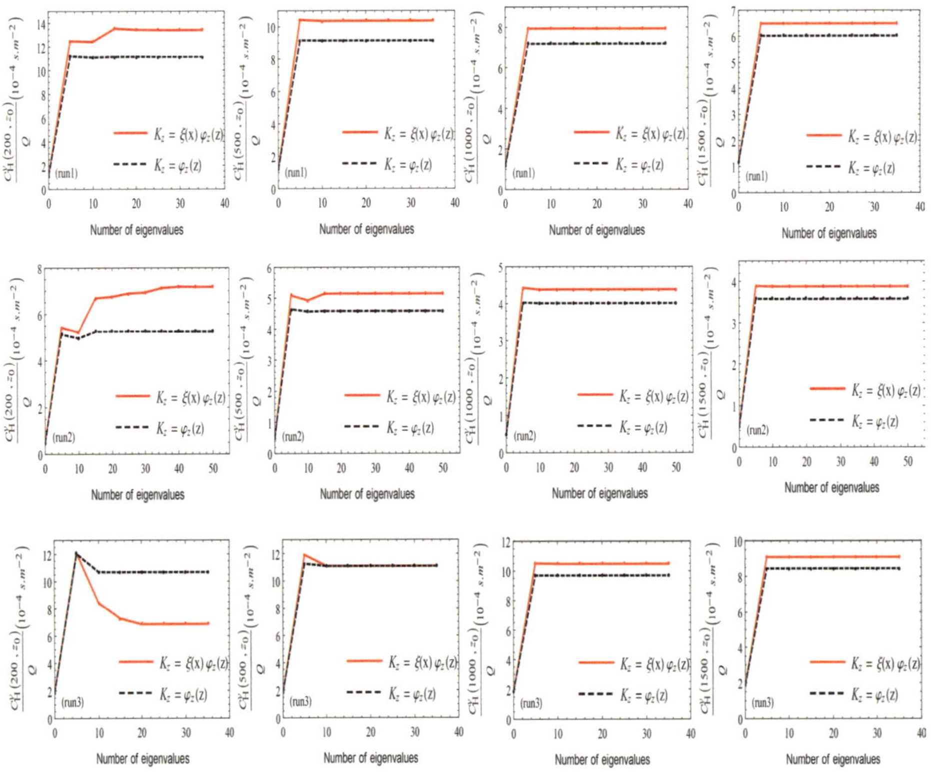

Figure 1.
Copenhagen experiments number 1, 2, and 3: Numerical convergence of the proposed solution for two vertical turbulence formulations as a function of the number of eigenvalues at four distances using four sub-layers.

The quality and performance of models are typically evaluated by comparing their predictions to observed (or measured) values. When observations are available, a common approach to evaluating model performance is to plot the predicted values against the observed values in a scatter diagram.

In this study, the performance of the models for the Copenhagen and Prairie Grass experiments was evaluated using scatter diagrams, which are shown in **Figures 4** and 5, respectively. **Figure 4** compares the predicted and observed crosswind-integrated concentrations for two different formulations of the vertical diffusivity parameter and two different numbers of sub-layers used to represent the atmospheric boundary layer (with $H = 2$ and $H = 4$), while **Figure 5** compares the predicted and observed ground-level crosswind integrated concentrations for two different numbers of sub-layers in the atmospheric boundary layer (again with $H = 2$ and $H = 4$).

The scatter diagrams show good agreement between the predicted and observed values, indicating that the models are well-parameterized. However, a visual inspection of the figures also reveals that the observed concentrations tended to be slightly higher than the predicted concentrations.

While scatter diagrams provide a useful visual evaluation, a more rigorous evaluation of model performance is achieved through statistical metrics. In this study, the statistical indices used to evaluate the models' performance are defined in the reference [23]. These statistical metrics provide a more detailed and quantitative

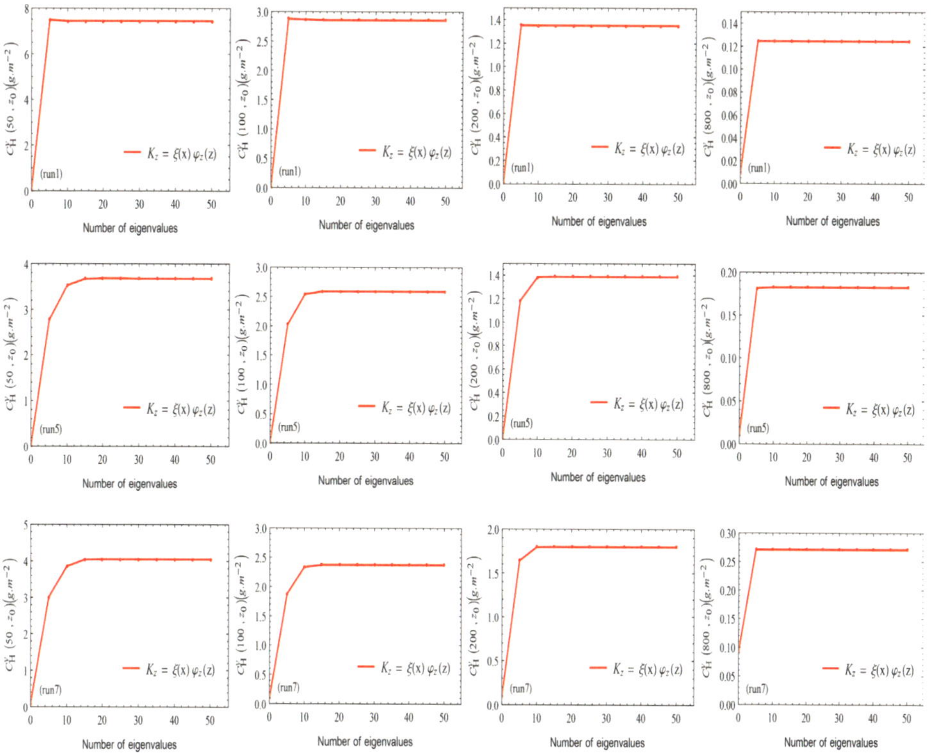

Figure 2.
Prairie grass experiment numbers 1, 5, and 7: Numerical convergence of the proposed solution as a function of the number of eigenvalues at four distances using four sub-layers.

assessment of the models' performance, which can be used to identify any areas of weakness or sources of error in the models. We defined the statistical indices used in our study as:

- Normalized Mean Square Error (NMSE): A measure of the accuracy of a model or prediction, calculated as the mean square error divided by the variance of the observed data:

$$NMSE = \frac{\overline{\left(c_o - c_p\right)^2}}{\overline{c_o} \times \overline{c_p}} \quad ,$$

- Mean Relative Square Error (MRSE): A measure of the accuracy of a model or prediction, calculated as the mean square error divided by the mean of the observed data squared:

$$MRSE = 4\overline{\left(\left(c_o - c_p\right)/\overline{\left(c_o + c_p\right)}\right)^2},$$

- Correlation Coefficient (COR): A measure of the linear relationship between two variables, usually the predicted and observed values of a model or prediction. It

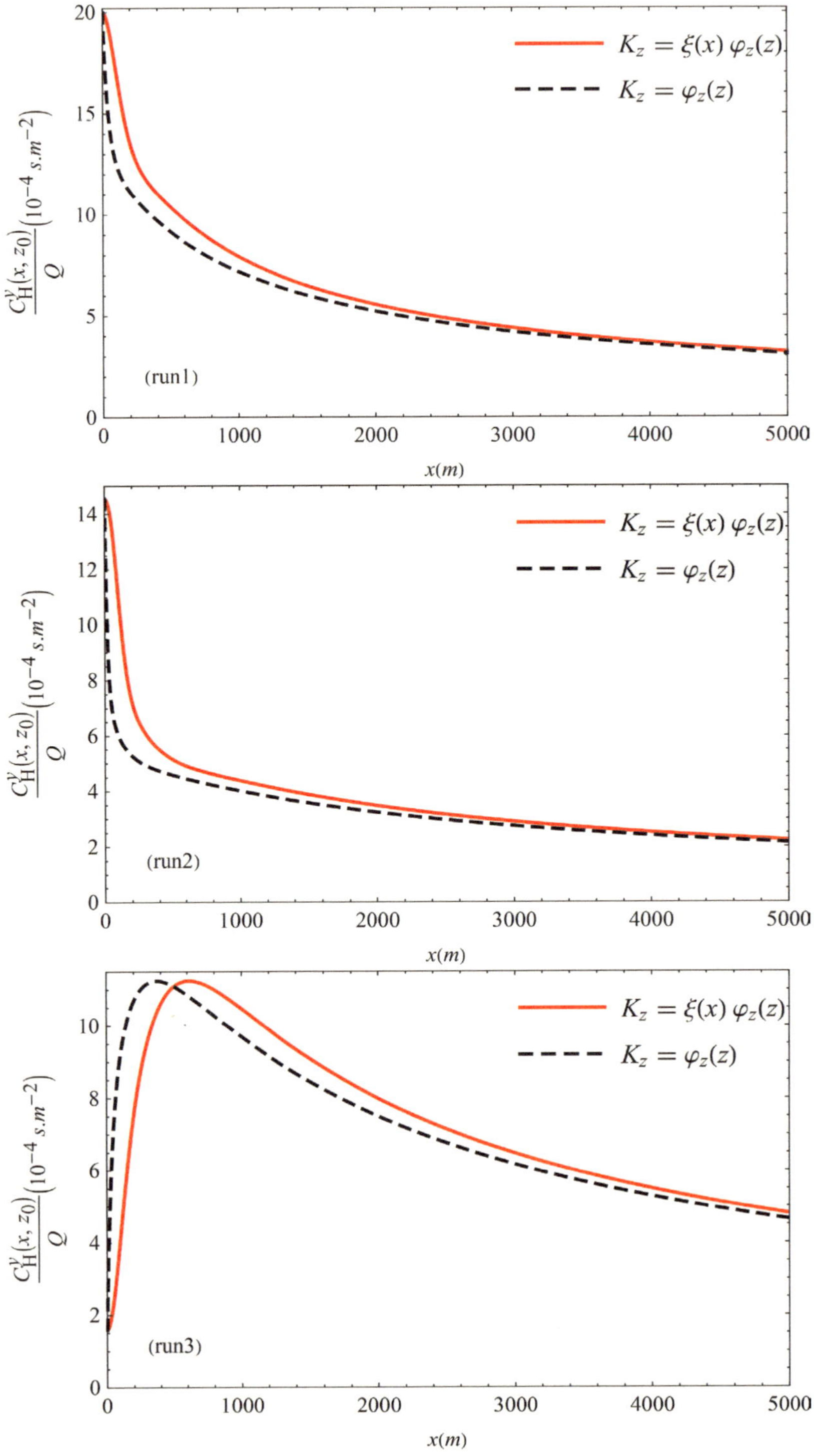

Figure 3.
Normalized concentration Eq. (24) as a function of distance using four sub-layers for Copenhagen experiment numbers 1, 2, and 3.

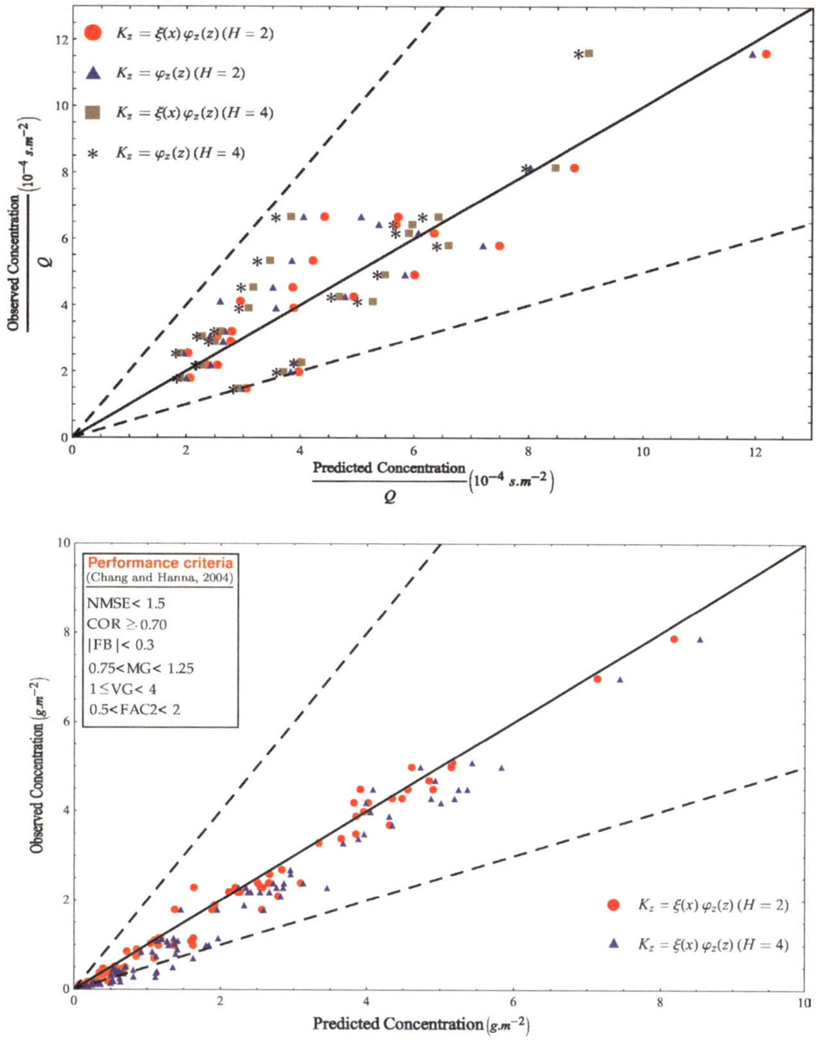

Figure 4.
Comparison of observed and predicted crosswind-integrated concentrations in Copenhagen (top panel) and prairie grass (bottom panel) experiments using scatter plots. The one-to-one line ($y = x$) and factor-of-two lines ($y = 0.5 \times x$ and $y = 2 \times x$) are shown.

ranges from -1 to 1, with 1 indicating a perfect positive correlation and -1 indicating a perfect negative correlation:

$$COR = \frac{\overline{\left(c_p - \overline{c_p}\right)\left(c_o - \overline{c_o}\right)}}{\left(\sigma_o \sigma_p\right)},$$

- Fractional Bias (FB): A measure of the systematic bias in a model or prediction, calculated as the difference between the mean predicted and observed values divided by the mean observed value:

$$FB = 2\,\frac{\left(\overline{c_o} - \overline{c_p}\right)}{\left(\overline{c_o} + \overline{c_p}\right)},$$

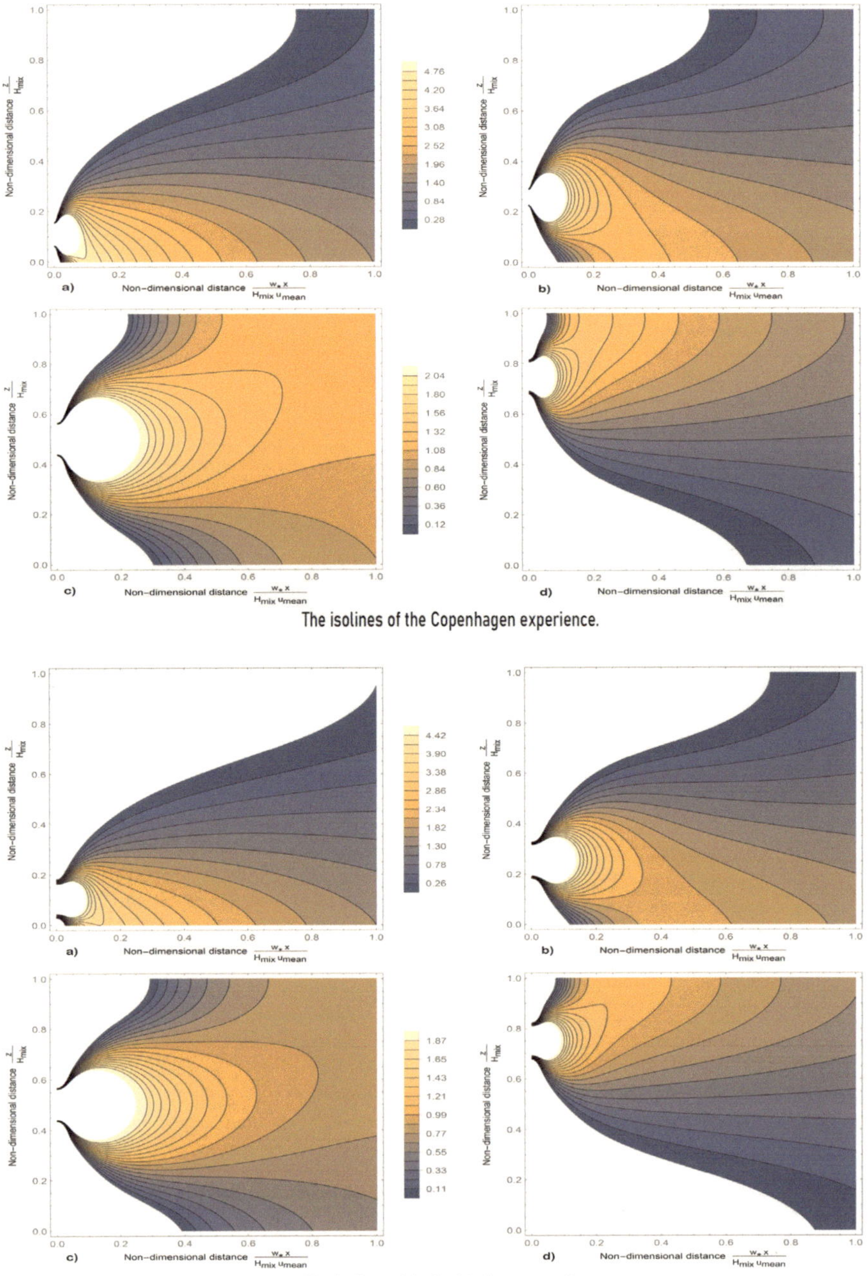

Figure 5.

Isolines of non-dimensional predicted crosswind-integrated concentration as a function of non-dimensional distance $\frac{w_* x}{u_{mean} H_{mix}}$ and non-dimensional depth $\frac{z}{H_{mix}}$ are presented for Copenhagen experiment number 2 (top panel) and prairie grass experiment number 7 (bottom panel) at different source heights: (a) $\frac{H_s}{H_{mix}} = 0.1$, (b) $\frac{H_s}{H_{mix}} = 0.25$, (c) $\frac{H_s}{H_{mix}} = 0.5$, and (d) $\frac{H_s}{H_{mix}} = 0.75$, using four sub-layers.

- Fractional Standard Deviation (FS): A measure of the variability in a model or prediction, calculated as the standard deviation of the predicted values divided by the mean observed value:

$$FS = 2\,\frac{(\sigma_o - \sigma_p)}{(\sigma_o + \sigma_p)}\ ,$$

- Mean Geometric Ratio (MG): A measure of the logarithmic bias in a model or prediction, calculated as the exponential of the mean of the differences between the logarithms of the predicted and observed values:

$$MG = \exp\left(\overline{\log(c_o)} - \overline{\log(c_p)}\right)\ ,$$

- Variance of Geometric Ratio (VG): A measure of the variability in the logarithmic bias of a model or prediction, calculated as the variance of the differences between the logarithms of the predicted and observed values:

$$VG = \exp\left[\overline{(\ln(c_o) - \ln(c_p))^2}\right]\ ,$$

- Factor of Two (FAC2): A measure of the accuracy of a numerical weather prediction model, defined as the distance between the observed and predicted position of a weather system divided by the radius of the weather system. If the distance is less than or equal to twice the radius, the forecast is considered accurate and has a FAC2 score of 1. If the distance is greater than twice the radius, the forecast is considered inaccurate and has a FAC2 score of less than 1:

$$0.5 \le (c_p/c_o) \le 2\ .$$

The satisfactory numerical results presented in **Figure 4** are supported by the values obtained from the statistical performance measures. Upon examining these values, it is observed that they mostly fall within the range of acceptable performance for the two proposed formulations for vertical turbulent diffusivity. However, it should be noted that the simulated concentrations for experiments conducted with Prairie Grass tend to slightly overestimate the measured quantities. Specifically, when considering the turbulent diffusivity that depends on both x and z, the model produces results that are relatively more promising than those obtained with other methods.

Figure 5 depicts the spatial distribution of pollutants for two different experiments —Copenhagen experiment number 2 and Prairie Grass experiment number 7, using contour lines that represent the adimensional transverse integrated concentration predicted based on the adimensional distance for experiment 2 and adimensional depth for experiment 7, where the adimensional distance for experiment 2 is defined as $w_* x/(u_{\text{mean}} H_{mix})$.

6. Conclusions

A solution has been developed for the 3-dimensional stationary state atmospheric diffusion equation, which incorporates more realistic formulations of wind velocity

	Copenhagen experiments							
Models	**NMSE**	**MRSE**	**COR**	**FB**	**FS**	**MG**	**VG**	**FAC2**
$K_z = \xi(x)\varphi_z(z)$ $(H = 2)$	0.0533	0.0533	0.8657	−0.0135	0.0018	0.9660	1.0885	1.096
$K_z = \varphi_z(z)$ $(H = 2)$	0.0660	0.0660	0.8498	0.0432	0.03897	1.0235	1.1007	1.055
$K_z = \xi(x)\varphi_z(z)$ $(H = 4)$	0.0768	0.0768	0.8308	0.0543	0.1686	1.0273	1.1022	0.9235
$K_z = \varphi_z(z)$ $(H = 4)$	0.0892	0.0890	0.8274	0.0973	0.2065	1.0726	1.1124	0.8882
[24] $(H = 2)$	0.08	—	0.86	−0.02	0.05	—	—	1.0
[24] $(H = 4)$	0.1	—	0.82	0.1	0.04	—	—	0.92
[25]	0.069	—	—	−0.009	0.051	0.996	1.055	1.009
	Prairie Grass experiments							
$K_z = \xi(x)\varphi_z(z)$ $(H = 2)$	0.0210	0.0209	0.9809	−0.0503	−0.0108	0.9018	1.0645	0.9807
$K_z = \xi(x)\varphi_z(z)$ $(H = 4)$	0.0613	0.0608	0.9763	−0.1809	- 0.0514	0.8616	1.3228	0.9373
[26]	0.25	—	0.92	0.03	0.20	—	—	0.68
[27]	0.04	—	0.96	−0.09	0.13	—	—	0.79

Table 1.
Statistical measures for the Copenhagen and Prairie grass experiments.

profiles and two formulations of vertical turbulent diffusivity. This solution was achieved through a combination of an appropriate auxiliary eigenvalue problem with mathematical induction, while a transcendental equation was developed to determine the eigenvalues for any number of subdomains. The advection-diffusion equation was solved by dividing the planetary boundary layer into discrete layers and assuming average values of turbulent diffusivity and wind velocity in each subdomain. The solution was evaluated using the Copenhagen and Prairie Grass experiments for two and four layers, and the numerical convergence of the solution was verified based on the number of eigenvalues. The results indicated good agreement between predicted and observed values, and most of the calculated statistical indices were within an acceptable range for model performance. This current model could be a promising approach for accurately predicting atmospheric pollutant dispersion and may also be applicable to other continuous flows (**Table 1**).

Author details

Mehdi Farhane* and Otmane Souhar
Laboratory of Modeling and Control of Stochastic and Deterministic Systems, Department of Mathematics, Chouaib Doukkali University, Faculty of Sciences El Jadida, Morocco

*Address all correspondence to: farhane.m@ucd.ac.ma

IntechOpen

© 2023 The Author(s). Licensee IntechOpen. This chapter is distributed under the terms of the Creative Commons Attribution License (http://creativecommons.org/licenses/by/3.0), which permits unrestricted use, distribution, and reproduction in any medium, provided the original work is properly cited.

References

[1] Moreira D, Vilhena M, Tirabassi T, Buske D, Cotta R. Near-source atmospheric pollutant dispersion using the new GILTT method. Atmospheric Environment. 2005;**39**:6289-6294 Available from: http://linkinghub.elsevier.com/retrieve/pii/S1352231005006023

[2] Buske D, Vilhena M, Moreira D, Tirabassi T. Simulation of pollutant dispersion for low wind conditions in stable and convective planetary boundary layer. Atmospheric Environment. 2007;**41**:5496-5501 Available from: http://linkinghub.elsevier.com/retrieve/pii/S1352231007003998

[3] Tirabassi, T., Buske, D., Moreira, D. and Vilhena, M. A two-dimensional solution of the advection–diffusion equation with dry deposition to the ground. Journal of Applied Meteorology and Climatology. 47, 2096-2104 (2008), ISBN: 1558-8432

[4] Vilhena M, Buske D, Degrazia G, Quadros R. An analytical model with temporal variable eddy diffusivity applied to contaminant dispersion in the atmospheric boundary layer. Physica A: Statistical Mechanics and its Applications. 2012;**391**:2576-2584 ISBN: 0378-4371

[5] Kumar P, Sharan M. An analytical model for dispersion of pollutants from a continuous source in the atmospheric boundary layer. Proceedings of the Royal Society A. 2010;**466**:383-406 Available from: http://royalsocietypublishing.org/doi/10.1098/rspa.2009.0394

[6] Vilhena M, Costa C, Moreira D, Tirabassi T. A semi-analytical solution for the three-dimensional advection–diffusion equation considering non-local turbulence closure. Atmospheric Research. 2008;**90**:63-69 Available from: http://linkinghub.elsevier.com/retrieve/pii/S0169809508000914

[7] Pérez Guerrero J, Pimentel L, Oliveira-Júnior J, Heilbron Filho P, Ulke A. A unified analytical solution of the steady-state atmospheric diffusion equation. Atmospheric Environment. 2012;**55**:201-212 Available from: http://linkinghub.elsevier.com/retrieve/pii/S1352231012002567

[8] Kumar P, Sharan M. An analytical dispersion model for sources in the atmospheric surface layer with dry deposition to the ground surface. Aerosol and Air Quality Research. 2016;**16**:1284-1293 Available comment: http://aaqr.org/articles/aaqr-15-09-oa-0549

[9] Deaves DM, Harris RI. Construction Industry Research and Information Association. A Mathematical Model of the Structure of Strong Winds. London: CIRIA; 1978. 49 p

[10] Mooney CJ, Wilson JD. Disagreements between gradient-diffusion and Lagrangian stochastic dispersion models, even for sources near the ground. Boundary-layer meteorology 1993;**64**:291-296

[11] Degrazia GA, Campos Velho HF, Carvalho JC. Nonlocal exchange coefficients for the convective boundary layer derived from spectral properties. Contributions to Atmospheric Physics. 1997;**70**

[12] Huang C. A theory of dispersion in turbulent shear flow. Atmospheric Environment. 1967;**13**:453-463 (1979,1). Available from: http://linkinghub.elsevier.com/retrieve/pii/0004698179901392

[13] Brown M, Pal Arya S, Snyder WP. Descriptors derived from a non-Gaussian

concentration model. Atmospheric Environment. 1997;**31**:183-189 Available from: http://linkinghub.elsevier.com/retrieve/pii/1352231096004876

[14] De Visscher A. Air Dispersion Modeling: Foundations and Applications. Somerset New Jersey: Wiley; 2014. http://www.books24x7.com/marc.asp?bookid=62593

[15] Pasquill F, Smith FB. Atmospheric diffusion: Study of the dispersion of windborne material from industrial and other sources. New York, NY, USA: John Wiley & Sons; 1983

[16] Mikhailov M, Ozisik M. Unified Analysis and Solutions of Heat and Mass Diffusion. New York: John Wiley and Sons Inc.; 1984 Available from: http://www.osti.gov/biblio/5949945

[17] Hanna SR, Briggs GA, Hosker JRP. Handbook on Atmospheric Diffusion: Prepared for the Office of Health and Environmental Research Office of Energy Research U.S. Department of Energy. Oak Ridge TN: Technical Information Center U.S. Dept. of Energy; 1982

[18] Obukhov A. Turbulence in an atmosphere with a non-uniform temperature. Boundary-Layer Meteorology. 1971;**2**:7-29 Available from: http://link.springer.com/10.1007/BF00718085

[19] Gryning S, Lyck E. Atmospheric dispersion from elevated sources in an urban area: Comparison between tracer experiments and model calculations. Journal of Applied Meteorology. 1984;**23**: 651-660 Available from: http://journals.ametsoc.org/doi/10.1175/1520-0450(1984)023%3C0651:ADFESI%3E2.0.CO;2

[20] Gryning S, Holtslag A, Irwin J, Sivertsen B. Applied dispersion modelling based on meteorological scaling parameters. Atmospheric Environment (1967). 1987;**21**(1):79-89 Available from: http://linkinghub.elsevier.com/retrieve/pii/0004698187902733

[21] Barad M. Project Prairie Grass: A Field Program in Diffusion, Vols. I and II of Geophysical Research Paper, 59, AFCRL-TR-58-235 (ASTIA Document No. AF-152572). Bedford: Air Force Cambridge Research Laboratories; 1958 Available from: http://www.harmo.org/jsirwin/

[22] Nieuwstadt F. An analytic solution of the time-dependent, one-dimensional diffusion equation in the atmospheric boundary layer. Atmospheric Environment (1967). 1980;**14**(1):1361-1364 Available from: http://linkinghub.elsevier.com/retrieve/pii/0004698180901547

[23] Chang J, Hanna S. Air quality model performance evaluation. Meteorology and Atmospheric Physics. 2004;**87** Available from: http://link.springer.com/10.1007/s00703-003-0070-7

[24] Marie E, Hubert B, Patrice E, Zarma A, Pierre O. A three-dimensional analytical solution for the study of air pollutant dispersion in a finite layer. Boundary-Layer Meteorology. 2015;**155**:289-300 Available from: http://link.springer.com/10.1007/s10546-014-9997-0

[25] Laaouaoucha D, Farhane M, Essaouini M, Souhar O. Analytical model for the two-dimensional advection-diffusion equation with the logarithmic wind profile in unstable conditions. International Journal of Environmental Science and Technology. 2022;**19**(7): 6825-6832. DOI: 10.1007/s13762-021-03554-1

[26] Moreira D, Tirabassi T, Vilhena M, Goulart A. A multi-layer model for pollutant dispersion with dry deposition

to the ground. Atmospheric Environment. 2010;**44**:1859-1865. Available from: http://linkinghub.elsevier.com/retrieve/pii/S1352231010001524

[27] Pérez Guerrero J, Pimentel L, Skaggs T. Analytical solution for the advection–dispersion transport equation in layered media. International Journal of Heat and Mass Transfer. 2013;**56**:274-282 Available from: http://linkinghub.elsevier.com/retrieve/pii/S0017931012007053

Chapter 6

A Review of Particle Removal Due to Thermophoretic Deposition

Yonggang Zhou, Mingzhou Yu and Zhandong Shi

Abstract

Thermophoretic deposition is an important technique for particle removal. The thermophoretic force of the particles under an appropriate temperature gradient can achieve a good particle removal effect. At present, there have been many studies on the deposition mechanism of ultrafine particles under the action of thermophoresis. In this chapter, the development history and current research status of the research on the thermophoretic deposition effect of ultrafine particles are summarized, and the future direction of thermophoretic deposition is proposed.

Keywords: ultrafine particle, deposition, thermophoresis, mathematical model, environmental gas pollution

1. Introduction

Fine particles play an important role in environmental gas pollution, which are smaller than coarse particles. These particles are main driver of urban haze formation, and also the main component of tobacco aerosol and kitchen fumes. They have an aerodynamic diameter of $2.5\mu m$ or less (PM2.5). The fine particles which are smaller than $0.1\mu m$ are referred to as ultrafine particles (PM0.1). These particles can often be suspended in the air and enter the respiratory tract of the human body along with the human body's respiration. Since the surface of fine particle is usually attached to harmful substances such as bacteria and heavy metals, it is easy to cause respiratory diseases and endanger human health after entering the human body [1]. Ebenstein et al. [2] found that combined atmospheric pollution has a significant impact on life expectancy and mortality from cardiopulmonary diseases. For evey $100\mu g/m^3$ increase in the concentration of air pollutants, the life expectancy of people under 5 years old is reduced by 1.5 years, and the life expectancy of people over 5 years old is reduced by 2.3 years. Cardiopulmonary diseases cause an increase of 79 deaths per 100,000 people. In 2016, a Polish cohort study of healthy school-age children aged 13–14 years found that for every quartile unit increase in PM1 concentration, forced vital capacity (FVC) and peak expiratory flow (PEF) decreased by 1.0 and 1.1%respectively [3].With the continuous development of manufacturing technology, in some electronic equipment and industrial products, the pollution caused by particle deposition affects both the yield and the use process. Some fine particles, after absorbing moisture in the air, will deposit on the surface of the equipment and form an electrolytic layer, which can have a corrosive effect on many metals. If the electrolyte penetrates into the protective layer of the wire

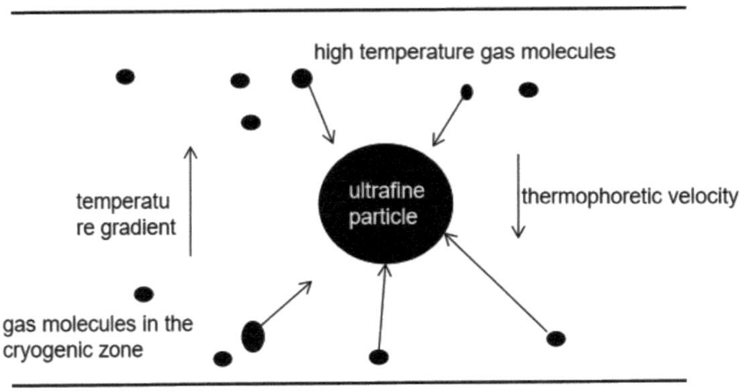

Figure 1.
Basic schematic diagram of thermophoresis.

to form corrosion points, an arc may be generated between the wire and the conductor to burn out the components.

The thermophoretic effect plays a very important role in the deposition of fine particles, so it is necessary to study the thermophoretic deposition of fine particles. In 2012, Ström and Sasic [4] studied the improvement of particle deposition efficiency in diesel or gasoline aftertreatment systems by thermophoresis. Their results show that for a standard monolithic channel with a gas-to-wall temperature difference of 200 K, deposition efficiencies of around 15% may be possible for all particle sizes. Some scholars have also applied it to the utilization of water resources. Dhanraj and Harishb et al. [5] designed a device to extract water from the air by thermophoresis and condensation. In the microgravity environment, thermophoresis plays a more critical role, and thermophoresis may be an important way of fine particle transfer. In the microgravity environment in space, it is of great significance to consider the influence of the thermophoresis mechanism in the aerospace field.

Although thermophoresis was discovered as early as the eighteenth century, its fundamental physical process was not proposed by Maxwell until 1879. In an area with a certain temperature difference, the gas molecules in the hot and cold areas continue to hit the particles. Since the gas molecules in the hot area have a large momentum, after hitting the particles, the macroscopically shows that the particles are subjected to force from the hot area to the particle. The phenomenon of movement of the cold zone, this force is the thermophoretic force [6]. Based on the formula of thermophoretic force, scientists derived formulas of thermophoretic deposition rate. In this chapter, the studies on both thermophoretic force and deposition rate due to thermophoresis are reviewed (**Figure 1**).

2. Thermophoretic deposition

In the temperature field with a certain temperature difference, the gas molecules and particles in the hot zone move to the hot zone again due to the reaction force after colliding with the particles. This phenomenon is called thermal glide. Maxwell (1879) and Reynolds (1880) analyzed the relationship between the flow field and the temperature field theoretically and experimentally [7] and determined the phenomenon of thermal slip. Maxwell proposed that the slip velocity of the gas is [8]:

$$\vec{v}_c = -\frac{3}{4}\nu \nabla lnT \tag{1}$$

where is the gas kinematic viscosity, T is the gas temperature.

2.1 Models of thermophoretic force

According to Maxwell's description and slip boundary conditions, Epstein selected spherical particles in the gas as the object, and the he solved the Navier-Stokes equation with the boundary conditions of the temperature gradient field considering the heat transfer heat balance.Finally,he first obtained the formula of thermophoretic force on the particles in the continuum region [9]:

$$F_T = \frac{9\pi\mu\nu d_p \overrightarrow{\nabla T}}{2T_o}\left(\frac{k_g}{k_p + 2k_g}\right) \tag{2}$$

here is fluid viscosity, is particle diameter, is fluid thermal conductivity coefficient, is particle thermal conductivity coefficient, is the average temperature of fluid near particle. Epstein's conclusion is in good agreement with the experimental results when and the thermal conductivity coefficient of the particles is low, but the theoretical value of this result is far from the experimental results when the thermal conductivity coefficient of the particles is high (L. [10]). Derjaguin and Storozhilova et al. [11]. pointed out that if high thermal conductivity particles such as sodium chloride are used, the actual thermophoretic force is two orders of magnitude higher than the theoretical [11].

In 1958, Waldmann [12] deduced the expression of the thermophoretic force of particles in the free molecular region in a single atomic gas based on the principle of molecular dynamics and the rigid body collision model between gas molecules and particles:

$$F_{th} = -\frac{16\sqrt{\pi}}{15}\frac{R^2 k_g}{\sqrt{\frac{2k_B T}{m_g}}}\nabla T \tag{3}$$

is Boltzmann constant, is gas molecular mass, is particle radius.

In 1962, based on Epstein's theory, Brock deduced the expressions of particle thermophoretic velocity and thermophoretic force applicable to the continuum region to the free-molecular region by applying the first-order slip-flow boundary conditions [13, 14]:

$$U_T = -\frac{2vC_s\left(\frac{k_g}{k_p} + 2C_t\frac{\lambda}{d_p}\right)\frac{(\nabla T)_x}{T_0}}{\left(1 + 4C_m\frac{\lambda}{d_p}\right)\left(1 + 2\frac{k_g}{k_p} + 4C_t\frac{\lambda}{d_p}\right)} \tag{4}$$

$$F_T = -\frac{12\pi\mu vRC_s\left(\frac{k_g}{k_p} + C_t\frac{\lambda}{R}\right)\frac{(\nabla T)_x}{T_0}}{\left(1 + 3C_m\frac{\lambda}{R}\right)\left(1 + 2\frac{k_g}{k_p} + 2C_t\frac{\lambda}{R}\right)} \tag{5}$$

where is the mean free path of gas molecules, $C_s(0.75)$, $C_m(1.14)$, and are dimensionless constants, which are thermal slip coefficient, viscous slip coefficient, and temperature jump coefficient, respectively. Although the result of brock still differs from the experimental value when the particles have a high thermal conductivity coefficient, it is already much smaller than Epstein's error.

Derjaguin [15] pointed out that Brock's assumption that the gas molecules always have the same velocity distribution before hitting the particle interface is inaccurate whose velocity distributions in the Knudsen layer vary with distance from the wall in the presence of a tangential temperature gradient. Through experiments, it is pointed out that its velocity formula (Eq. 4) does not conform to the experimental results at medium and large particle sizes ($0.3–0.6\ \mu m$).

In 1980, Talbot [10] used laser-Doppler velocimeter (LDV) to measure the thickness of the particle void region produced by the particle velocity distribution in the laminar boundary layer at the heated wall under the action of thermophoretic force. According to the experimental results and the BGK model in the particles, it is found that the reason for the large error of the Brock thermophoretic force formula (Eq. 4) is the problem of the thermal slip coefficient C_s, and the value of is corrected to 1.17. Talbot also proposed the expression of the thermophoresis coefficient k_{th}. Although the revised thermophoretic force expression is still not perfect, it has been widely used.

Cha, McCoy [16] and Wood [17] deduced a thermophoretic calculation model with good applicability in a wide range of Knudsen numbers. Although the theoretical calculation results and the experimental data reported by other scholars at that time have small numerical errors, they are in good agreement with the overall trend. After correcting the correlation coefficient, the calculated result is in good agreement with the result of Talbot's formula when $k_n < 3$. The corrected expression is:

$$F_{th} = 1.15\frac{Kn}{4\sqrt{2}\alpha\left(1+\frac{\pi_1}{2}Kn\right)}\cdot\left[1-\exp\left(-\frac{\alpha}{Kn}\right)\right]\cdot\left(\frac{4}{3\pi}\phi\pi_1 Kn\right)^{0.5}\frac{k_B}{d_m^2}\nabla T d_p^2 \quad (6)$$

$$\alpha = 0.22\left[\frac{\frac{\pi}{6}\phi}{\left(1+\frac{\pi_1}{2}Kn\right)}\right]^{\frac{1}{2}} \quad (7)$$

$$\pi_1 = \frac{0.18\left(\frac{36}{\pi}\right)}{\left[\frac{(2-S_n+S_t)4}{\pi}+S_n\right]} \quad (8)$$

where is the equivalent diameter of the channel, C_v is the constant volume-specific heat capacity, is the conventional momentum adjustment coefficient, and is the tangential momentum adjustment coefficient.

In 2003, Li and Wang [18, 19] deduced the theoretical calculation formula of the thermophoretic force on nanoparticles in the free molecular region based on the non-rigid collision model:

$$\boldsymbol{F}_{T,\mathrm{Li}} = -\frac{8}{3}\sqrt{\frac{2\pi m_r}{k_B T}}\kappa R^2 \nabla T\left(\frac{6}{5}\Omega^{(1,2)^*}-\Omega^{(1,1)^*}\right) \quad (9)$$

In the formula, is the mass of particles, is thermal conductivity of gases, $m_r = m_g m_p/(m_g+m_p)$, is the reduced mass of gas molecules and particles; $\Omega^{(1,1)^*}$ and $\Omega^{(1,2)^*}$ are the dimensionless collision integral, for rigid body collision,

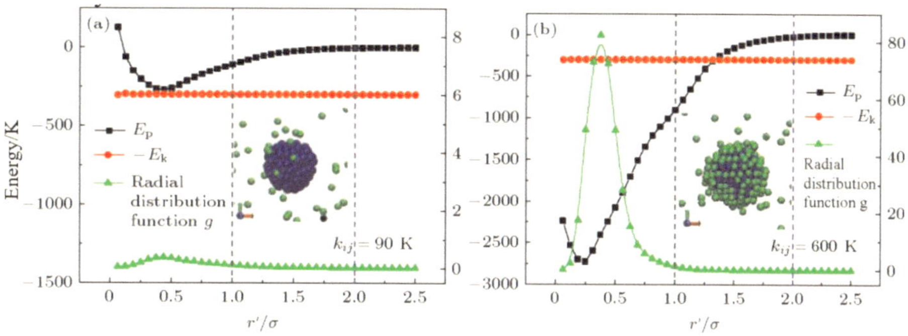

Figure 2.
The potential energy Ep, kinetic energy Ek and density functiong on the surface of nanoparticles Cui et al. [20].

$\Omega^{(1,1)^*} = \Omega^{(1,2)^*} = 1$. Eq. (3) is a special case of Eq. (9). But Li and Wang's theoretical model has not been experimentally demonstrated. Due to the insufficiency of nano-scale particle measurement technology, it is often difficult to study the corresponding thermophoresis phenomenon in experiments. Cui et al. [20] used molecular dynamics simulation to verify the theoretical calculation model of Waldmann [12] and Li [18, 19] found that when the particle size was reduced to the nanometer scale, Waldmann's calculation model is no longer applicable due to the enhancement of the non-rigid collision effect, and the theoretical model established by Li and Wang after considering the non-rigid collision effect is closer to the molecular dynamics simulation results. At high gas-solid bonding strength, gas molecules are easily adsorbed on nanoparticles, which changes the true particle size of the particles (**Figure 2**), resulting in a certain error in the correction formulas of Li and Wang [18, 19]. Cui et al. [20] corrected the particle size error and the results were basically consistent with the simulated values.

In terms of thermophoresis measurement technology, it is mainly divided into particle group measurement and single particle measurement. In the measurement using the particle group, the gas stream with particles enters a vacuum chamber with a temperature gradient in the vertical direction. Calculations are often inaccurate. Particle size can be determined when individual particles are measured, so it tends to be more precise. Li and Davis [21] used electrodynamic balance (EDB) to measure individual particles and compared with previous theories and obtained better data result (**Figure 3**).

2.2 Models of thermophoretic deposition efficiency

In the calculation of the deposition efficiency of thermophoresis, many scholars have established their expressions. Since the theoretical calculation models of various scholars under laminar flow conditions can be well in line with the experimental or simulated values, the calculation model of thermophoretic deposition efficiency under turbulent flow conditions is mainly introduced here.

In 1969 Byers and Calver [22] measured the influence of flow parameters and particle collector size on deposition efficiency and established the thermophoretic deposition efficiency expression:

$$\eta_L = 1 - \exp\left(-\frac{\rho C_p f \,\mathrm{Re}_D}{4Dh}\frac{K_{th}v(T_e - T_w)}{\overline{T}}\left(1 - \exp\left(\frac{-4hL}{u_m \rho C_p D}\right)\right)\right) \quad (10)$$

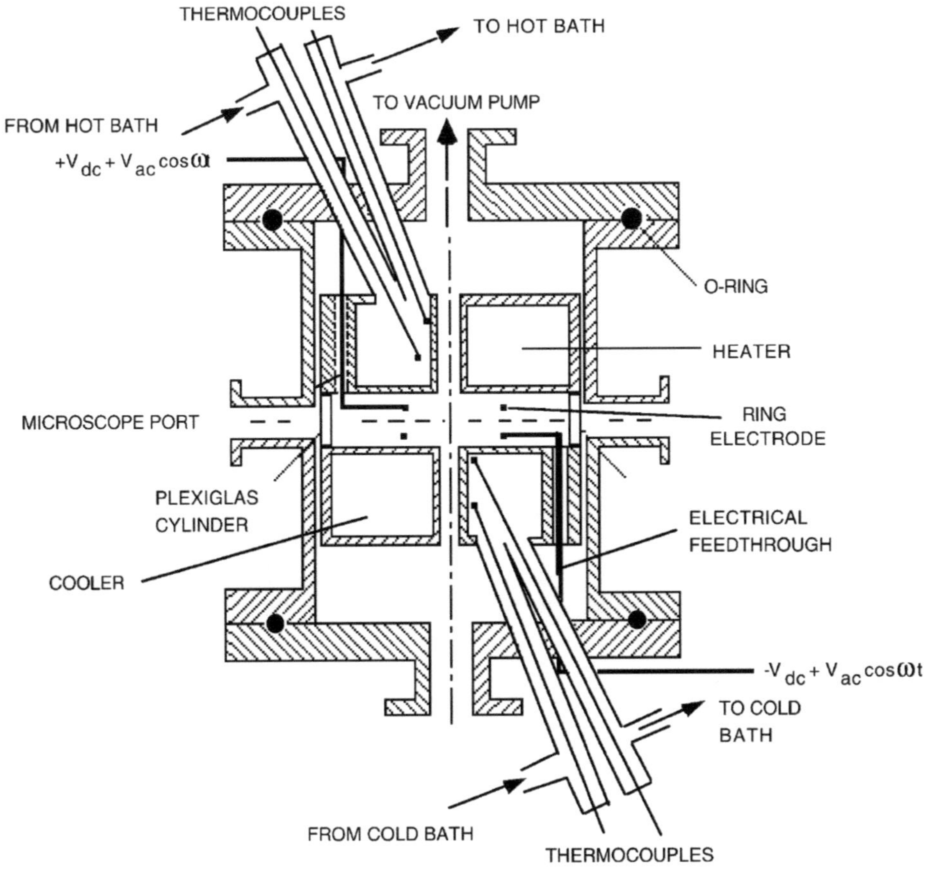

Figure 3.
A cross-section of the electrodynamic balance and vacuum chamber used by Li and Davis for thermophoretic force measurements [21].

where is the specific heat capacity of the gas at constant pressure, T_e and are the fluid inlet and tube wall temperatures respectively, K_{th} is thermophoretic coefficient, is the average temperature, $h$is the convective heat transfer coefficient, is the tube length, is the tube diameter, is the gas density, f is the coefficient of friction, is the average radial velocity.

In 1974, Nishio experimentally determined the deposition of aerosols on the temperature gradient along the length of the heat exchanger tube wall,

$$\eta_L = 1 - \exp\left(-\frac{\rho C_p K_{th} v (T_e - T_w)}{k_g \overline{T}}\left(1 - \exp\left(\frac{-4hL}{u_m \rho C_p D}\right)\right)\right) \tag{11}$$

Batchelor and Shen [23] used the similarity method to analyze the deposition rate as the pipe length by thermophoretic effects in flow over plates, cylinders, and rotating bodies,

$$\eta_L = \mathrm{Pr}\mathrm{K}_{th}\left(\frac{T_e - T_w}{T_e}\right)\left(1 + (1 - \mathrm{Pr}K_{th})\left(\frac{T_e - T_w}{T_e}\right)\right) \tag{12}$$

Where is Prandtl number of air.

In 1998, Romay et al. [24] conducted experiments in turbulent pipelines, respectively measured the influence of inlet flow velocity, flow rate, particle size, and inlet fluid temperature on the deposition efficiency of the pipeline, and compared them with the existing theoretical model Eqs. (10) and (11) of deposition efficiency at that time. For comparison, an expression for the deposition efficiency of thermophoresis in a turbulent tube is derived:

$$\eta_L = 1 - \left[\frac{T_w + (T_e - T_w)\exp\left(-\frac{\pi DhL}{\rho Q C_p}\right)}{T_e}\right]^{P_r K_{th}} \tag{13}$$

Q is the volume flow (**Figure 4**).

The above one-dimensional expressions are all derived under specific assumptions, and cannot be well applied to general engineering specifications. In 2005, Housiadas and Drossinos [25] developed a two-dimensional model including radial section effects after establishing the one-dimensional long-tube deposition efficiency expression suitable for laminar and turbulent flow and carried out extensive verification. Its one-dimensional expression is:

$$\eta_L = 1 - \left(\frac{\theta^*}{1+\theta^*}\right)^{\mathrm{Pr}K_{th}} = 1 - \left(\frac{T_w}{T_e}\right)^{\mathrm{Pr}K_{th}} \tag{14}$$

There are many studies on the calculation of key parameters under the thermophoretic effect. Because the preconditions selected by each scholar are different, the establishment of the final expression is also different. Early studies on thermophoretic deposition effects mostly focused on the mechanism. In recent years,

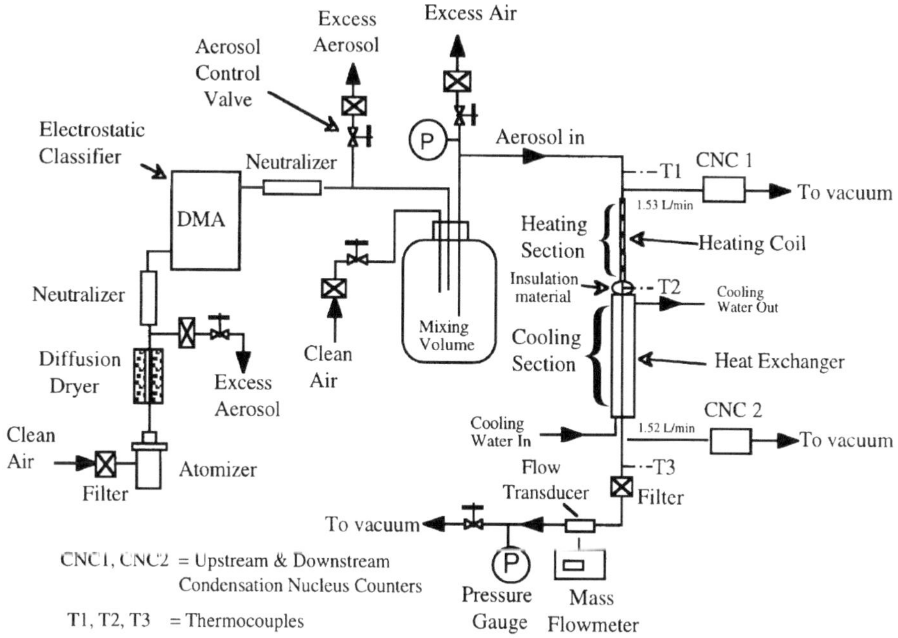

Figure 4.
Schematic diagram of the experimental apparatus by Romay et al. [24].

with the maturity of numerical simulation and particle detection technologies, more scholars have begun to consider the performance of thermophoretic deposition in different scenarios. The development direction is gradually diversified.

3. Current status of particle thermophoretic deposition

At present, there are three main ways to study the thermophoretic effect: theoretical research, numerical simulation, and experimental research. Researchers deduce new empirical expressions based on existing theories through numerical simulations or experiments under selected specific experimental conditions. The force characteristics and motion trajectory of the particles are analyzed, and the particle motion model of the thermophoresis effect under different conditions is established. The main work is divided into three categories. One is the influence of particle shape on thermophoretic effect, the other is the study of thermophoretic deposition effect of particles in pipes or small spaces, and the changes in particle concentration distribution or deposition law caused by thermophoretic effect in indoor and outdoor temperature fields. The following is an overview of the results of the main research groups currently working on both types of work.

3.1 Research on thermophoretic deposition in pipes and tiny spaces

Thermophoretic effects in pipes or tiny spaces are a hot topic in current thermophoretic research. The researchers selected specific flow field conditions and pipeline structures and established corresponding particle motion models based on existing theories through experiments and numerical simulations. The research team of Lin and Tsai of National Chiao Tung University used the critical trajectory method to study the effect of developing flow in a circular tube on the deposition efficiency of thermophoretic particles [26]. Through theoretical and numerical analysis, a dimensionless equation for calculating thermophoretic deposition efficiency under laminar flow conditions is established. The research results show that the inlet section where the velocity and temperature are both developing is more conducive to the formation of thermophoretic sedimentation of particles due to the existence of a relatively large temperature gradient, resulting in higher thermophoretic deposition efficiency. They determined the effect of particle diffusion and particle electrostatic charge induced deposition on thermophoretic deposition efficiency in laminar and turbulent tubes [27] (**Figure 5**). It is found that the deposition efficiency caused by the electrostatic charge of particles is comparable to the thermophoretic deposition efficiency when the thermophoretic efficiency is usually lower than 10% in their experiments (**Figure 6**), so this effect should be excluded when calculating the thermophoretic deposition efficiency to obtain accurate experimental data. Even for particles in Boltzmann charge balance, the deposition efficiency of the particle electrostatic charge has a considerable effect compared to the thermophoretic deposition efficiency. In addition, Tsai et al. also studied the inhibition process of particle thermophoretic deposition through the tube wall when the tube wall temperature exceeds the gas temperature in a circular tube [28]. The particle transport equations of convection, diffusion, and thermophoresis were numerically solved, and the particle concentration distribution and deposition laws were obtained. The results show that for all particle sizes, the particle deposition rate decreases with increasing tube wall temperature and gas flow rate. When the tube wall is heated to a certain temperature slightly above the gas temperature, particle deposition is completely suppressed. And given dimensionless deposition parameters, an empirical expression is established to predict the dimensionless temperature difference required for zero deposition in laminar flow tubes (**Figures 5** and **6**).

The Lee's team studied the effect of thermophoresis on the deposition rate of particles above the wafer in a clean room environment [29, 30]. Using the statistical

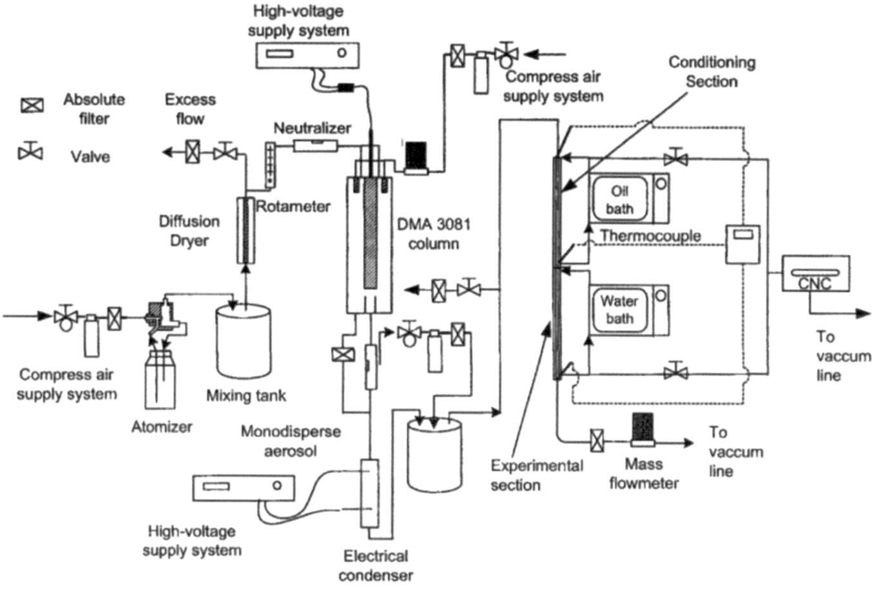

Figure 5.
Schematic diagram of the experimental apparatus by Lin and Tsai [27].

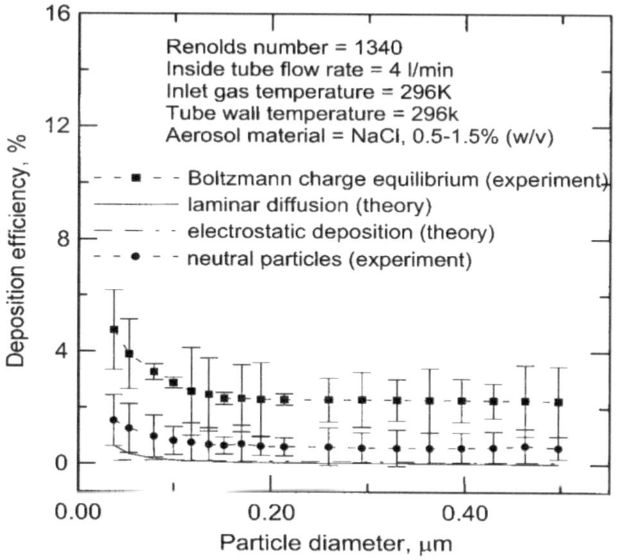

Figure 6.
Comparison of experimental deposition efficiencies (nonthermophoretic) and theoretical predictions of diffusional and electrostatic deposition under laminar flow conditions (Re = 1340).

Lagrangian particle tracking (SLPT) model, under the condition of parallel airflow, particle deposition rates above individual wafers were measured. The law of particle deposition velocity as a function of temperature difference (temperature difference between plane and ambient air), particle density, and parallel airflow velocity is summarized. With and without thermophoresis, some numerical simulation results are as follows.The results show that with the increase of particle density, the particle deposition velocity decreases sharply with the increase of particle size, and the increase of airflow velocity also leads to the increase of particle deposition velocity (**Figures** 7 and 8).

Research on the deposition effect of particles in pipes or tiny spaces is a very popular direction in thermophoretic research both at home and abroad. Many scholars choose different parameters or channel and space conditions to establish corresponding particle motion models under thermophoretic effects. Yu et al. [31] carried out numerical simulation and analysis on the influence of thermophoretic force in MOCVD horizontal reactor on the concentration distribution of reaction precursors during deposition. For the TMGa molecules in the MOCVD reactor, the calculation formulas of the thermophoretic force, thermophoretic velocity, and diffusion velocity were deduced; Ho et al. [32] studied the effect of thermophoresis on particle deposition rate in mixed convection on vertically corrugated plates, using a cubic spline method, combining dimensionless variables, Prandtl transforms, and parabolic transform to obtain the final result.The results show that the smaller the particle size, the greater the influence of electrophoresis and thermal swimming, the greater the influence of temperature gradient and electric field on the deposition rate of particles.

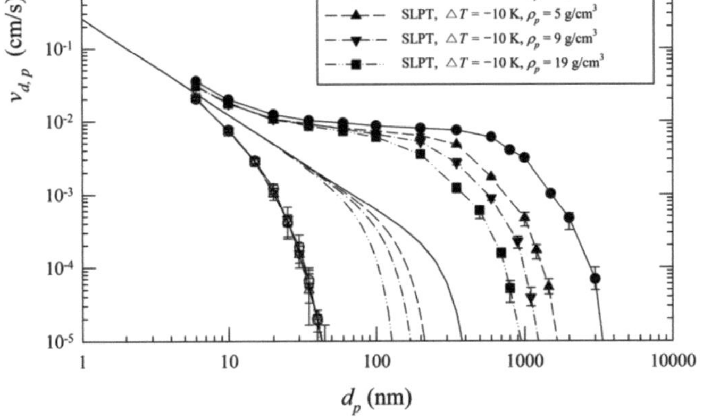

Figure 7.
Effect of particle density.

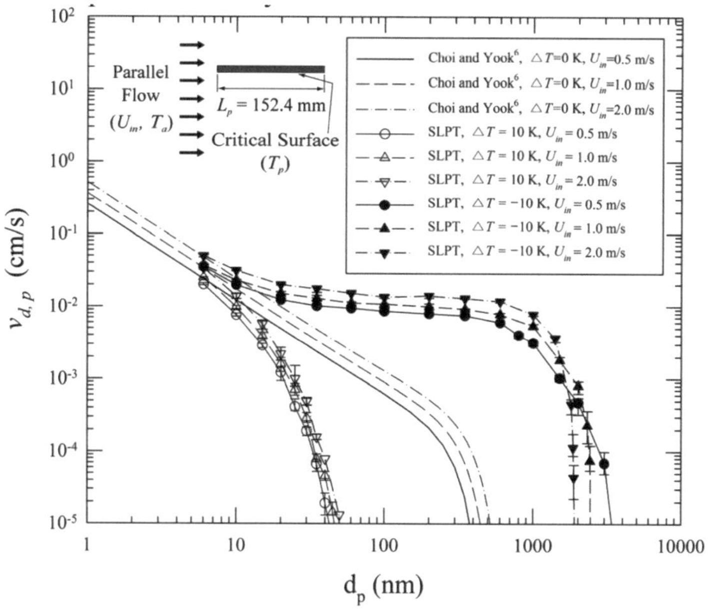

Figure 8.
Effect of airflow velocity.

3.2 Research on thermophoretic deposition under indoor and outdoor temperature fields

Compared with the first type of work, there are few international studies on particle thermophoretic deposition in indoor temperature fields, and the experimental conditions chosen by various scholars are quite different.

The team of Xu and Chen [33, 34] proposed a zero-equation model to simulate the three-dimensional distribution of indoor air velocity, temperature, and pollutant concentration. This method assumes that turbulent viscosity is a function of length scale and local average velocity. This new computational model is much faster than the standard model. A two-layer model was then used to predict the flow. The model adopts the single equation model for the near-wall region and the standard model for the outer wall region, which improves the calculation efficiency again. Some scholars have also applied it to the utilization of water resources.

Lai [35] used two chambers to simulate indoor and outdoor conditions and the crack module to simulate wall cracks to study the effect of thermophoresis on particle penetration cracks. By simulating summer and winter conditions in temperate climate regions, it is found that the penetration ratio of particles from indoor to outdoor in winter conditions is significantly higher than that in summer and isothermal conditions when the particle size is less than 100 μm. And the effect of temperature on the particle penetration ratio decreases with the increase of particle size.

3.3 Study on the influence of thermophoresis effect on particle deposition between rotating disks

Due to the geometry of the disk, the fluid flow under the action of the rotating disk is widely used in many engineering fields. In the past decade, with the development of mechanical technology, the effect of thermophoretic deposition on the fluid flow on the rotating disk has gradually become a hot topic.

Khan and Mahmood [36] investigated the effect of thermophoretic deposition on MHD flow of Oldroyd-B nanofluids between radiatively stretched disks. They used the homotopy analysis method to solve the transformed ordinary differential equations, and analyzed the convergence of the obtained series solutions. Through the analysis of the obtained results, it was found that the fluid temperature and the nanoparticle concentration decreased with the increase of the thermophoretic velocity parameter value. But it will increase with the increase of thermophoretic diffusion parameters. Hafeez et al. [37] also studied the thermophoretic deposition of particles in Oldroyd-B fluid between rotating disks, using von Karman similarity variables to convert partial differential equations into dimensionless ordinary differential equations, and with the help of Maple Numerical format (BVP-Midrich technique) to obtain numerical solutions. The results show that the axial thermophoretic velocity of the particles between the discs increases with the increase of the relative thermophoretic coefficient; the slope of the particle concentration growth curve decreases with the increase of the thermophoretic coefficient; the inward axial thermophoretic deposition velocity (Local Stanton number) increases with the increase of the thermophoresis coefficient (**Figure 9**).

Since the transformed ordinary differential equations are coupled and nonlinear, it is difficult to obtain closed-form solutions. Therefore, in recent years, many scholars have adopted the shooting technique and used the Runge-Kutta integral scheme to solve the ordinary differential equations obtained after similarity transformation. M.S. Alam et al. [38] used this method to study the deposition mechanism of micron-sized particles caused by thermophoresis during transient forced convection heat and mass transfer on an impermeable rotating disk with a surface temperature lower than that of the surrounding fluid. Doh and Muthtamilselvan [39] studied the effect of a rotating disk in a uniform electromagnetic field on the deposition of thermophoretic particles during unsteady heat and mass transfer in forced convection in a micropolar fluid. Gowda et al. [40] investigated the deposition of thermophoretic particles in a mixed nanofluidic flow suspended by ferrite nanoparticles. For different fluid or particle objects, most scholars have obtained relatively consistent results on the changing trends of related thermophoretic parameters in the thermophoretic deposition between rotating discs.

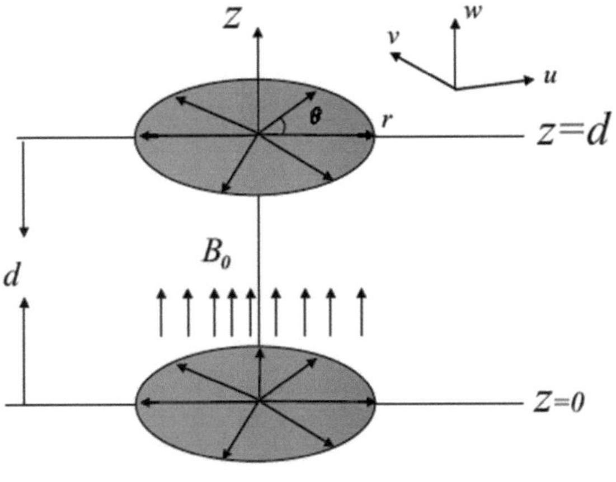

Figure 9.
A physical sketch of the problem by Khan and Mahmood [36].

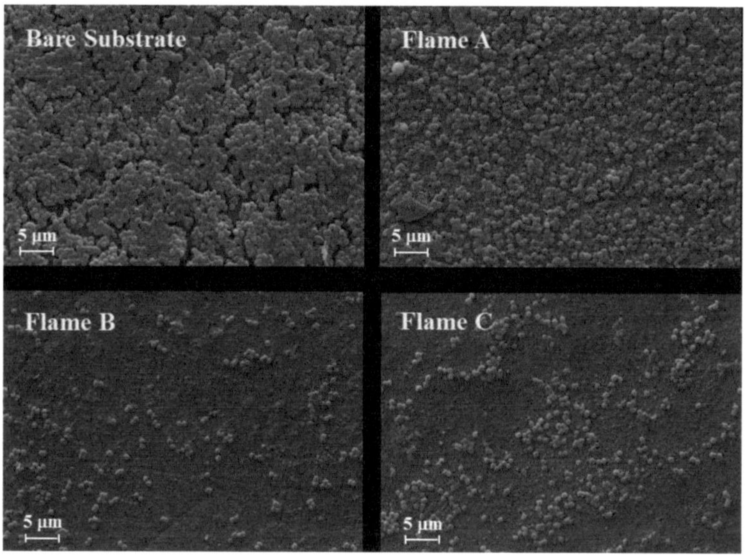

Figure 10.
SEM microphotographs of Staphylococcus aureus *biofilm on uncoated aluminum substrates and* TiO_2 *coated aluminum substrates for the three different flame conditions.*

3.4 Research on thermophoretic deposition technology in other works

In addition to the above main types of work, thermophoretic deposition technology is also widely used in other aspects. Shi and Zhao [41] considered the deposition velocity of particles on the human body surface under Brownian and turbulent diffusion, gravitational settling and thermophoresis. The results show that for particles below 1 μm, thermophoresis is the main deposition mechanism. Brugière and Gensdarmes et al. [42] designed a radial flow thermophoretic velocity analyzer device. By developing the transfer function of the device they designed, the instrument can directly measure the particle velocity in the temperature field with high resolution. Effective thermophoretic velocity, eliminating the need to build a model to calculate the thermophoretic behavior of particles. De Falco et al. [43] achieved a highly controllable and tunable technique for the production of nanostructured TiO_2 coating films on aluminum substrates by combining aerosol flame synthesis and direct thermophoretic deposition. Self-cleaning and self-disinfecting coating materials with near superhydrophilicity and high antibacterial activity can be prepared. As shown, the number of *Staphylococcus aureus* was significantly reduced on the substrate with the titanium dioxide layer (**Figure 10**).

4. Conclusion and outlook

Nowadays, the international research on the thermophoretic deposition effect has been very extensive, and a lot of achievements have been achieved. However, there are still many deficiencies and limitations in both the theoretical understanding of the thermophoretic mechanism and the application of the thermophoretic deposition effect in practical work.

Regarding the understanding of the thermophoretic mechanism, there have been relatively mature research results in the thermophoretic particle motion model in the continuum region and free molecular region, but some scholars have extended their theory to the transition region but found that the results are not ideal; Most of the studies are carried out under the condition that the thermal conductivity of particles is similar to the thermal conductivity of gas. When the thermal conductivity of particles is higher, the experimental results have a large deviation from the theoretical calculation values; In terms of particle shape, there are many studies on spherical particles, and the theoretical and experimental data are in good agreement. However, there are few studies on non-spherical particles, and most of the particles are non-spherical in reality. Errors introduced by thermophoretic force measurements on non-spherical particles may affect the data accuracy of the entire experiment.

In terms of experiments, the experimental conditions selected by various scholars are quite different, and it is difficult to make horizontal comparisons; Thermophoretic force is a short-range force. To obtain higher particle deposition efficiency in practical engineering applications, it is necessary to design an effective removal device structure; In the process of particle deposition, there is often the problem of particle resuspension, so the subsequent treatment of the deposited particles is also very important. At present, many studies have not considered the effects of particle agglomeration and fragmentation. In practical engineering applications, the characteristics of the flow field where the particles are located are usually much more complicated than the conditions selected in the experiment, so there are still many problems to be explored to widely apply thermophoresis technology to practical work.

Acknowledgements

The authors thank the Key R & D plan of Zhejiang Province (2019C03097), National Natural Science Foundation of China (11872353) and Natural Science Foundation of China of Zhejiang Province (LZ22A020004) for their support.

Nomenclature

T	gas temperature
F_T	thermophoretic force
d_p	particle diameter
k_g	fluid thermal conductivity coefficient
k_p	particle thermal conductivity coefficient
T_o	average temperature of fluid near particle
Kn	Knudsen numberv
k_B	Boltzmann constant
m_g	gas molecular mass
R	particle radius
d_m	equivalent diameter of the channel
C_v	constant volume-specific heat capacity
S_n	conventional momentum adjustment coefficient
S_t	tangential momentum adjustment coefficient
m_p	mass of particles

m_r	reduced mass of gas molecules and particles
$\Omega^{(1,1)^*}, \Omega^{(1,2)^*}$	dimensionless collision integral, for rigid body collision
C_p	specific heat capacity of gas
T_e	fluid inlet temperatures
T_w	tube wall temperatures
$\overline{T}$	average temperature
h	convective heat transfer coefficient
L	tube length
D	tube diameter
f	coefficient of friction
u_m	average radial velocity
Pr	Prandtl number of air
K_th	thermophoretic coefficient
Q	volume flow

Greek letters

$\vec{v}_c$	slip velocity of gas
ν	gas kinematic viscosity
μ	fluid viscosity
λ	mean free path of gas molecules
κ	thermal conductivity of gases
ρ	gas density
η_L	thermophoretic deposition efficiency

Abbreviations

FVC	forced vital capacity
PEF	peak expiratory flow
LDV	laser-Doppler velocimeter
SLPT	statistical Lagrangian particle tracking
EDB	electrodynamic balance

Author details

Yongggang Zhou[1], Mingzhou Yu[1*] and Zhandong Shi[2]

1 Laboratory of Aerosol Science and Technology, China Jiliang University, Hangzhou, Zhejiang

2 Zhengzhou Tobacco Research of CNTC, Zhengzhou, China

*Address all correspondence to: mzyu@cjlu.edu.cn

IntechOpen

© 2023 The Author(s). Licensee IntechOpen. This chapter is distributed under the terms of the Creative Commons Attribution License (http://creativecommons.org/licenses/by/3.0), which permits unrestricted use, distribution, and reproduction in any medium, provided the original work is properly cited.

References

[1] Zhao M, Zhou C, Chan T, Tu C, Liu Y, Yu M. Assessment of COVID-19 aerosol transmission in a university campus food environment using a numerical method. Geoscience Frontiers. 2022;**2022**:101353

[2] Ebenstein A, Fan M, Greenstone M, He G, Yin P, Zhou M. Growth, pollution, and life expectancy: China from 1991–2012. American Economic Review. 2015; **105**(5):226-231

[3] Zwozdziak A, Sówka I, Willak-Janc E, Zwozdziak J, Kwiecińska K, Balińska-Miśkiewicz W. Influence of PM1 and PM2.5 on lung function parameters in healthy schoolchildren—A panel study. Environmental Science and Pollution Research. 2016;**23**(23): 23892-23901

[4] Ström H, Sasic S. The role of thermophoresis in trapping of diesel and gasoline particulate matter. Catalysis Today. 2012;**188**(1):14-23

[5] Harish S, Dhanraj RG, Loganathan GB. An efficiency study on water extraction from air using thermophoresis method. IOP Conference Series: Materials Science and Engineering. 2019;**574**(1):012003

[6] Zheng F. Thermophoresis of spherical and non-spherical particles: A review of theories and experiments. Advances in Colloid and Interface Science. 2002; **97**(1–3):255-278

[7] Bakanov SP. Thermophoresis in gases at small knudsen numbers. Aerosol Science and Technology. 1991;**15**(2): 77-92

[8] Kennard EH. Kinetic theory of gases. International Series in Physics. 1938; **57**(39):901

[9] Epstein, Von Paul S. Zur Theorie Des Radiometers. 1927;**54**(7-8):537-563

[10] Talbot L. Thermophoresis of particles in a heated boundary layer. Mechanical Engineering. 1980;**1980**: 646-655

[11] Derjaguin BV, Storozhilova AI, Rabinovich YI. Experimental verification of the theory of thermophoresis of aerosol particles. Journal of Colloid and Interface Science. 1966;**21**(1):35-58

[12] Waldmann L. Die Boltzmann-Gleichung Für Gase Aus Spinteilchen. Zeitschrift Fur Naturforschung - Section A Journal of Physical Sciences. 1958; **13**(8):609-620

[13] Brock JR. On the theory of thermal forces acting on aerosol particles. Journal of Colloid Science. 1962;**17**(8):768-780

[14] Brock JR. The thermal force in the transition region. Journal of Colloid and Interface Science. 1967;**23**(3): 448-452

[15] Derjaguin BV, Yalamov Y. Theory of thermophoresis of large aerosol particles. Journal of Colloid Science. 1965;**20**(6): 555-570

[16] Cha CY, McCoy BJ. Thermal force on aerosol particles. Physics of Fluids. 1974; **17**(7):1376-1380

[17] Wood NB. The mass transfer of particles and acid vapor to cooled surfaces. 1981;**76**:6-93

[18] Li Z, Wang H. Drag force, diffusion coefficient, and electric mobility of small particles. I. Theory applicable to the free-molecule regime. Physical Review E - Statistical Physics, Plasmas,

Fluids, and Related Interdisciplinary Topics. 2003a;**68**(6):1-9

[19] Li Z, Wang H. Drag force, diffusion coefficient, and electric mobility of small particles. II. Application. Physical Review E – Statistical Physics, Plasmas, Fluids, and Related Interdisciplinary Topics. 2003b;**68**(6):061206

[20] Cui J, Jun Jie S, Wang J, Xia GD, Li ZG. Thermophoretic force on nanoparticles in free molecule regime. Wuli Xuebao/Acta Physica Sinica. 2021; **70**(5):1-9

[21] Li W, James E. Measurement of the Thermophoretic Force by Electrodynamic Levitation: Microspheres in air 1995;**26**(7): 1063-1083

[22] Byers RL, Calvert S. Particle deposition from turbulent streams by means of thermal force. Industrial & Engineering Chemistry Fundamentals. 1969;**8**(4):646-655

[23] Batchelor GK, Shen C. Thermophoretic deposition of particles in gas flowing over cold surfaces. Journal of Colloid and Interface Science. 1985; **107**(1):21-37

[24] Romay FJ, Takagaki SS, Pui DYH, Liu BYH. Thermophoretic deposition of aerosol particles in turbulent pipe flow. Journal of Aerosol Science. 1998;**29**(8): 943-959

[25] Housiadas C, Drossinos Y. Thermophoretic deposition in tube flow. Aerosol Science and Technology. 2005; **39**(4):304-318

[26] Lin JS, Tsai CJ. Thermophoretic deposition efficiency in a cylindrical tube taking into account developing flow at the entrance region. Journal of Aerosol Science. 2003;**34**(5):569-583

[27] Tsai CJ, Lin JS, Aggarwal SG, Chen DR. Thermophoretic deposition of particles in laminar and turbulent tube flows. Aerosol Science and Technology. 2004;**38**(2):131-139

[28] Lin JS, Tsai CJ, Chang CP. Suppression of particle deposition in tube flow by thermophoresis. Journal of Aerosol Science. 2004;**35**(10):1235-1250

[29] Kim W-K, Lee S-C, Yook S-J. Numerical investigation of Thermophoretic effect on particulate contamination of an inverted flat surface in a parallel airflow. Journal of the Electrochemical Society. 2011;**158**(10): H1010

[30] Woo SH, Lee SC, Yook SJ. Statistical Lagrangian particle tracking approach to investigate the effect of thermophoresis on particle deposition onto a face-up flat surface in a parallel airflow. Journal of Aerosol Science. 2012;**44**:1-10

[31] Yu HQ, Zuo R, Chen JS, Peng XX. Analysis and numerical simulation of precursor concentration distribution on the influence of Thermophoretic force on deposition process in horizontal MOCVD reactor. Rengong Jingti Xuebao/Journal of Synthetic Crystals. 2011;**40**(4):1033-1038

[32] Ho PY, Chen CK, Huang KH. Combined effects of thermophoresis and electrophoresis on particle deposition in mixed convection flow onto a vertical wavy plate. International Communications in Heat and Mass Transfer. 2019;**101**(January):116-121

[33] Chen Q, Weiran X. A zero-equation turbulence model for indoor airflow simulation. Energy and Buildings. 1998; **28**(2):137-144

[34] Xu W, Chen Q. A two-layer turbulence model for simulating indoor

airflow - Part I. Model development. Energy and Buildings. 2001;**33**(6): 613-625

[35] Lai ACK. An experimental study of indoor and outdoor concentrations of fine particles through nonisothermal cracks. Aerosol Science and Technology. 2013;**47**(9):1009-1016

[36] Khan N, Mahmood T. Thermophoresis particle deposition and internal heat generation on MHD flow of an Oldroyd-B Nanofluid between radiative stretching disks. Journal of Molecular Liquids. 2016;**216**: 571-582

[37] Hafeez A, Khan M, Ahmed J. Oldroyd-B fluid flow over a rotating disk subject to soret–dufour effects and thermophoresis particle deposition. Proceedings of the Institution of Mechanical Engineers, Part C: Journal of Mechanical Engineering Science. 2021; **235**(13):2408-2415

[38] Alam MS, Chapal Hossain SM, Rahman MM. Transient thermophoretic particle deposition on forced convective heat and mass transfer flow due to a rotating disk. Ain Shams Engineering Journal. 2016;**7**(1):441-452

[39] Doh DH, Muthtamilselvan M. Thermophoretic particle deposition on magnetohydrodynamic flow of micropolar fluid due to a rotating disk. International Journal of Mechanical Sciences. 2017;**130**(April):350-359

[40] Punith Gowda RJ, Naveen Kumar R, Aldalbahi A, Alibek I, Prasannakumara BC, Rahimi-Gorji M, et al. Thermophoretic particle deposition in time-dependent flow of hybrid nanofluid over rotating and vertically upward/ downward moving disk. Surfaces and Interfaces. 2021;**22** (October 2020):100864

[41] Shi S, Zhao B. Deposition of indoor airborne particles onto human body surfaces: A modeling analysis and manikin-based experimental study. Aerosol Science and Technology. 2013; **47**(12):1363-1373

[42] Brugière E, Gensdarmes F, Ouf FX, Yon J, Coppalle A, Boulaud D. Design and performance of a new device for the study of thermophoresis: The radial flow thermophoretic analyser. Journal of Aerosol Science. 2013;**61**:1-12

[43] Falco G, D, Ciardiello R, Commodo M, Del Gaudio P, Minutolo P, Porta A, et al. TIO2 nanoparticle coatings with advanced antibacterial and hydrophilic properties prepared by flame aerosol synthesis and thermophoretic deposition. Surface and Coatings Technology. 2018;**349**:830-837

Chapter 7

Indoor Air Pollution

Abebaw Addisu

Abstract

Indoor air pollution becomes a public health hazard across the world. It originates from different sources such as the use of unclean fuel in developing countries for cooking, heating, and lighting purposes. Their use results in incomplete combustion. Carbon monoxide and other toxic gases are the primary result of incomplete combustion and can cause respiratory tract problems. Children and women who spent a large portion of their time indoors are the most vulnerable subpopulation.

Keywords: indoor pollutants, particulate matter, carbon monoxide, unclean fuel, respiratory tract infections

1. Introduction

Indoor air pollution, also known as household air pollution is a state of increased concentration of air pollutants within and around a building or structure at a degree high enough to harm human beings and the ecosystem at large.

Air pollutants can arise from indoor activities like cooking, heating, and lighting using unclear fuels such as firewood, charcoal, animal dung, and kerosene [1]. The incomplete combustion of those fuels results in the production of carbon monoxide, nitrogen oxides, and sulfur oxides [2]. Other indoor activities like the application of insecticides, pesticides, air conditioners, and tobacco smoke can also contribute a large share of indoor particulate matter concentration [3]. Particulate matter is the complex mixture of dust, soot, smoke, and liquid droplet that suspends in the air in a small amount that can be inhalable.

Exposure to indoor air pollutants makes susceptible individuals breathe in those pollutants and activate as much as 15 sensory nerve receptors in the respiratory tract system that leads to reflex mechanism [4]. Pollutants that have a diameter of less than 5 micro-meter are candidates to reach alveoli by passing against physiological and mechanical barriers of the respiratory tract system [5]. After inhalation and availability in distal respiratory receptors, the pollutants cause acute respiratory infection most commonly pneumonia among young children and exacerbate asthma, chronic lung disease, and cancer among adults [6].

Although air pollution becomes a global concern that affects the population of the world in all spheres, indoor air pollution is a particular health risk for low-income countries [7]. Groups of sub-population such as infants, women, elderly, person with chronic disease, and urban residents who spent a large proportion of their time indoors, are at higher risk when compared to the general population [8].

IntechOpen

Special locations including schools, homes, and workplaces (wood processing industries and cement factories) are sites with greater concerns. Ventilation and meteorological parameters such as temperature and humidity play a crucial role in the concentration of indoor air pollutants.

Indoor air pollution becomes a public health concern that results in a broad array of health problems ranging from reversible to irreversible damage to the body system. So, it is important to discover the source from which pollutants arise and develop skills on how to measure and monitor indoor pollutants to combat against its adverse health effect.

2. Sources of indoor air pollution

Sources of indoor air pollution can be broadly classified into two as:

1. Anthropogenic (Man-made)

2. Natural

Anthropogenic sources of indoor air pollution are sources that originate from human activities for different purposes.

2.1 Environmental Tobacco Smoke (ETS)

It is also known as secondhand smoke exhaled by smokers. ETS contains a mixture of more than 7357 compounds, of which at least 40 of them are carcinogenic [9]. The only source of ETS is the combustion of tobacco products at home, vehicles, and office [10]. The combustion of tobacco products emits chemicals in the form of particulate matter ($PM_{2.5}$ and PM_{10}) and vapor form.

2.1.1 Combustion of products

Human beings throughout the world utilize biomass fuels to various degrees with the greatest proportion belonging to developing and economically disadvantageous nations. Solid fuels such as firewood, animal dung, and charcoal are the primary source of energy that accounts for 90% of consumption in sub-Saharan Africa for the purpose of cooking, heating, and lightening [11]. Other sources like gas water heater, gas cloth dryer, kerosene/gas space heaters are combustible products.

The combustion of those products can release gaseous pollutants most importantly carbon oxides (CO, CO_2), nitrogen dioxide, and sulfur dioxide.

Carbon monoxide is asphyxiant gas that has a greater affinity to hemoglobin (oxygen transporter for tissue) than oxygen does [2]. This affinity results in the formation of carboxyhemoglobin and disrupts oxygen transport.

2.1.2 Volatile organic compounds (VOCs)

VOCs are compounds that emit gas at room temperature from certain solids or liquids [12]. Nowadays, pesticides are largely used in agriculture and food processing; emit small particles that can easily pollute indoor air quality. Moreover, air conditioners, hair spray building materials constitute a greater proportion of VOCs. Stationary materials such as copy machines and printers, adhesives, and permanent markers can also contribute to VOCs.

Natural sources of indoor air pollution occur without human intervention.

2.1.3 Biological contaminants

Animal dander, molds, and dust mites are the common sources of biological contamination that give rise to particulate concentration and resultant indoor air pollution [4].

There are conditions that encourage the growth of those biological contaminants and their presence in the air:

- High relative humidity: that encourages dust mite population and hastens fungal growth on the damp surface.

- Inadequate exhaust for bathroom/kitchen generated moisture.

- Appliances such as humidifiers, dehumidifiers, and drip pans under cooling coins (refrigerator) support bacterial and fungal growth.

2.1.4 Heavy metals (airborne lead and mercury vapor)

Although nowadays items become lead-free, older housing and furniture are coated with lead. Additionally, art and craft materials, automobile radiators, and solders are the main sources of lead that lead to pollutant concentration.

Mercury vapor: nowadays new paints have emerged as a concern to contain a high level of mercury in water-based paints in the form of phenyl mercuric acetate (PMA). The sprinkling of mercury by some ethnic or religious groups for ritual activities. Moreover, it is used for herbal medicine and botanic shops [12].

2.1.5 Asbestosis and radon

Both are known carcinogenic agents that arise from structural fireproofing and acoustic installation in floor and ceiling tiles. With the advancement of age, asbestos-containing materials become damaged and disintegrate which gives rise to microscopic fiber into the air which increases the concentration of indoor air pollutants.

Radon: naturally occurring colorless, odorless, and tasteless radioactive gas that arises from the decay of radium or uranium. Earth-derived building materials and underground water especially, well water from private supplies are the major sources of radon pollution in households [8].

2.2 Measurements of indoor air pollutants

2.2.1 Particulate matter

Particulate matter (PM) is a term used to describe extremely small particles and liquid droplets in the atmosphere suspended in a gaseous medium. It is a complex mixture of particles including dust, soot, smoke, and liquid droplets which are generally small enough to be inhaled [13].

Based on aerodynamic diameters, particulate matter can be broadly classified into two

1. Coarse particulate matter (PM_{10}): particle matters having a diameter of greater than 2.5 micrometers and less than 10 micrometers.

2. Fine particulate matter ($PM_{2.5}$): particle matters of 2.5 micrometers and less.

The concentration of particulate matter confined in the indoor environment can be expressed in two ways as:

- Particle number concentrations (count per cm^3, L, m^3)

- Particle mass concentrations ($\mu g/m^3$ or mg/m^3)

Different instruments can measure the concentrations of particulate matter and provide either an average concentration for the sampling period or real-time monitoring concentrations [3].

Four major techniques can be employed to measure the concentration of PM.

2.2.2 Gravimetric principle (filter-based samplers)

The basic principle is collecting target particles on a filter. Air is allowed to be drawn with a certain flow rate with the help of a pump. Filters are going to be measured before (unloaded) and after (loaded) sampling keeping temperature and relative humidity to the standard condition. Simply, the difference between the two measurements gives us the mass of captured particulate matter. The mass concentration is then determined by dividing the collected mass by the known volume of air drawn through the filter by the pump. Accurate time-average mass concentration could be obtained for the period of sampling.

Impaction or centrifugal force is crucial for removing particles larger than the target aerodynamic diameter from the air stream.

2.2.3 Microbalance principle (tapered element oscillating microbalance or TEOM)

The principle behind this method is just putting a filter on the hollow element which is oscillating with its pattern when air passes through the filter. This oscillating frequency varies over time and the change of frequency is the measure of PM concentration build-up. It produces continuous data.

2.2.4 Beta-ray attenuation principle

After collecting the target particles on the filter, the mass of PM is determined by directing beta-rays from radioactive sources through the filter and the particles on it. The measure of PM mass is determined by the beam crossing the filter.

2.2.5 Light-scattering principle (photometry)

Usually, portable devices suck sample air through a chamber. Inside the chamber, sample air is provided with a narrow beam of artificial light. A photodetector measures the amount of light scattering via sampled air-containing particles. These devices are factory calibrated, contain own pump and storage. It can provide continuous data.

Generally, particulate matter (PM) has to be measured at the different locations inside the living house such as the kitchen, where most of the emission comes from fuels, living room, and outdoors for comparison. Measurement of PM should be:

- Approximately 1 meter away from the edge of the combustion zone (from the stove approximates the edge of the active cooking area.

- At the height of the sitting position (1 meter) above the floor which is a breathing zone for sitting women while cooking and children.

- At least 1.5 meters horizontally away from windows and doors, where possible [14].

The outdoor particulate matter has to be measured at the sampling location of height about 2 meters from the ground level, a minimum of 2 meters away from obstructions like the wall of the house, 10 meters from trees, 5 meters from the chimney, and 5 meters from the edge of the nearest traffic lane.

WHO set a standard for air quality monitoring for different pollutants, among which PM is the most significant. According to the organization, the standard provides the maximum and interim target values. Interim targets are proposed as incremental steps in the reduction of air pollution and are intended for use in areas where pollution is high 1] (**Table 1**).

2.3 Carbon monoxide (CO)

It is a colorless, odorless, and tasteless gas that is poorly soluble in water. Relatively high carbon monoxide levels have been measured inside homes with faulty or unvented combustion appliances, particularly if the appliances have been used in poorly ventilated rooms.

2.3.1 Analytical method

The reference method for the measurement of carbon monoxide concentration is based on the absorption of infrared radiation (IR) by the gas in a non-dispersive photometer [15]. CO absorbs IR radiation maximally at a wavelength of 4.7 micrometers (μm), which is in a spectral range where few other atmospheric species absorb to interfere with accurate quantification.

Pollutants	Measurement	Averaging time	Interim Target				AQG level
			1	2	3	4	
$PM_{2.5}$	μg/m^3	Annual	35	25	15	10	5
		24-hour	75	50	37.5	25	15
PM_{10}	μg/m^3	Annual	70	50	30	20	15
		24-hour	150	100	75	50	45

Table 1.
WHO standard concentration of particulate matter at different stages of measurement.

Carbon dioxide and water vapor can make major interference with the measurement of carbon monoxide. Removal of water vapor from the sample air is necessary to avoid positive interferences in the determination of CO concentration and is achieved by a permeation tube or drier that selectively removes water vapor from the sample gas without removing CO.

The concentration of CO can be determined using the Beer–Lambert law which relates the concentration of an absorbing species to the degree of light attenuation. The analytical method is suitable for stable installations at fixed-site monitoring stations.

Currently, potable carbon monoxide individual-level exposure measurements become familiar. Measurements of personal exposures are based on the electrochemical reactions between carbon monoxide and deionized water, which are detected by specially designed sensors.

The following guideline values (ppm values rounded) and periods of time-weighted average exposures have been determined in such a way that the COHb level of 2.5% is not exceeded, even when a normal subject engages in light or moderate exercise:

- 100 mg/m^3 (90 ppm) for 15 minutes
- 60 mg/m^3 (50 ppm) for 30 minutes
- 30 mg/m^3 (25 ppm) for 1 hour
- 10 mg/m^3 (10 ppm) for 8 hours

2.4 Estimation and measurement of VOC emissions

VOC can be measured at stationary level and individual concentration (personal level) [16].

2.4.1 Total VOC concentration measurement techniques

- Flame ionization detectors (FID)
- Catalytic oxidation and non-dispersive infrared absorption
- Photoionization detection (PID)

2.4.2 Individual VOC substance concentration measurement techniques

- Gas Chromatography (GC)
- Non-Dispersive Infrared Spectrometry (NDIR)
- Fourier Transform Infrared absorption (FTIR)

2.5 Health effects of indoor air pollutants

Indoor air pollutants that are emitted from different sources occupy the living and get entrance to our internal body systems usually through inhalation.

Although the target organs may vary from one pollutant to another, the most common are the lungs, eyes, and nervous system. It is important to understand the kinetics and metabolism of pollutants for better knowledge of their effects on the human body system.

2.6 Health effect of carbon monoxide

Kinetics of carbon monoxide starts with inhalation of the pollutant and diffuses to the alveolar membrane of the lung. During a gas exchange that takes place in capillaries, the CO gas can be exchanged in the same way oxygen does. After its bioavailability in plasma, quickly binds with hemoglobin (Hb) at rate 200–300 times more than oxygen [15]. It can equivocally compete for hemoglobin binding sites with oxygen. Unlike oxygen, the bond is stronger and lasts longer. The formation of carboxyhemoglobin (COHb) makes the transport of oxygen difficult for the body system and results in a state of hypoxia. The primarily affected body organs with reduced oxygen delivery are the myocardium (muscle of the heart), brain, skeletal muscles that employ exercise, and developing fetus. At low levels, symptoms of CO exposure include fatigue, headaches, and dizziness, but in higher concentrations, it can lead to impaired vision, disturbed coordination, nausea, and eventually death.

2.7 Health effects of biological agents at particular level

It is important to broadly classify the health effects into three categories as follows:

1. Infection: direct invasion of tissues and organs by particulate-level biological agents like dust mites, molds, and animal dander causes pathogenic infections.

2. Hypersensitivity disease: inhalation and absorption of certain pollutants can evoke and activate the specific immune system.

3. Toxicosis: biologically produced toxins such as mycotoxins (fungal metabolites) can cause direct toxic effects.

2.8 Health effect of pesticides

Nowadays, pesticides become input for effective agricultural practice and are widely used across the urban and rural communities. Inhalation, ingestion, and skin absorption are the common exposure routes of particles in pesticides. Pesticide exposure is associated with adverse health risks, including short-term skin and eye irritation, dizziness, headaches, and nausea; and long-term chronic impacts, such as cancer, asthma, and diabetes.

A case–control study has shown the association of breast cancer with exposure to certain pesticides in agricultural sectors [1].

2.9 Health effect of heavy metals

Heavy metals like mercury vapor, lead, chromium, and cadmium residue are known for their bioaccumulation and biomagnification properties. Bioaccumulation is the kinetic property of the chemicals in which metabolism and excretion of certain elements are difficult and result in accumulation in specific tissues and organs of the human body system. An increase in the concentration of those heavy metals when they came across the complex food chain is termed as biomagnification.

Studies have proved that Mutagenic and carcinogenic effects are common health effects related to heavy metals [4]. Moreover, brain damage, respiratory illnesses, cardiovascular disease, and deaths can be attributed as a result of it.

2.10 Health effect of particulate matter (PM)

Exposure to particulate matter is considered as a public health hazard. It can primarily attack the lung followed by the heart and eye. It has acute and chronic effects.

Acute exposure to those particles can cause:

- Persistent cough
- Shortness of breath
- Chest tightness
- Eye irritation
- Irregular heartbeat
- Nonfatal heart attacks

Chronic exposure to the particles can cause:

- Decreased lung function
- Recurrent respiratory infections among children
- Premature death in people with heart or lung disease
- Aggravated asthma

A cross-sectional study has shown the association of acute respiratory infections with particulate matter concentration ($PM_{2.5}$ and PM_{10}) from biomass fuel exposure in an indoor environment [14].

Strategic approaches to reduce indoor air pollution.

Indoor air pollution become a public health issues that need urgent interventions to save the life of millions of people and keep the environment from damage. Different approaches have been presented at international, regional, and national levels. Implementing those approaches is vital to coping the existing situation.

2.11 Energy sources development

It is impossible to achieve indoor air quality without the provision of clean energy. A large proportion of the world relies on unclean fuel that results in incomplete combustion. Energy preference largely depends on the economic power of the nation. Policies and strategies that enable a nation to use natural resources like water for hydroelectric power, which is environmentally friendly and helps transform energy to clean have to be implemented in developing countries.

2.12 Pollution prevention principles

As a human being we can prevent the occurrence of pollution before it happens and reduce consecutively its impact if it occurs. Pollution can be prevented at the source, path, and receiver (usually individuals). Prevention at source means avoiding pollution occurrence in the first place by either eliminating, banning, or restricting the usage of hazardous chemicals at commercial levels. Pollution prevention at the path includes proper ventilation of living rooms which is expressed in terms of window floor ratio. Finally, prevention at the receiver can be using personal protective equipment like wearing masks where the concentration of pollutants is higher.

2.13 Taxation

Selective taxations are crucial for tobacco products in order to minimize environmental tobacco smoke at household level. Additionally, environmental taxation on leaded petrol has brought great influence and was successful [17].

2.14 Information dissemination, education and training

Adequate information has to be disseminated for policymakers to design appropriate strategies to address the issue. Moreover, designers, building, and construction professionals have to be informed about the raw materials they use as construction input.

3. Conclusion

As a result of human activity for the sake of earning a living and urbanization, humans exploit nature and use different raw materials that can end up in the generation of pollutants that affect the health of human beings. The effect is significant among occupants of the household that spend a large portion of time indoors usually younger children and women. Regular monitoring of pollutants using real-time monitors is crucial to understand the degree of risk and designing appropriate prevention strategies.

Author details

Abebaw Addisu
Department of Environmental Health, College of Medicine and Health Science, Mizan-Tepi University, Mizan, Ethiopia

*Address all correspondence to: abebawaddisu11@gmail.com

IntechOpen

© 2023 The Author(s). Licensee IntechOpen. This chapter is distributed under the terms of the Creative Commons Attribution License (http://creativecommons.org/licenses/by/3.0), which permits unrestricted use, distribution, and reproduction in any medium, provided the original work is properly cited.

References

[1] C. C. Attribution-noncommercial-sharealike and C. C. By-nc-sa. Compendium of WHO and other UN guidance on health and environment Chapter 2. Air pollution, vol. 2022, pp. 0-28, 2022

[2] Colome & Samet. Indoor Air Pollution: An Introduction for Health Professionals. 1995

[3] Salvador S, Salvador E. Overview of Particle Air Pollution Air Quality Communication Workshop. San Salvador. 2012

[4] Van Tran V, Park D, Lee Y. Indoor Air Pollution, Related Human Diseases, and Recent Trends in the Control and Improvement of Indoor Air Quality. International Journal of Environmental Research and Public Health. 2020. pp. 1-27. DOI: 10.3390/ijerph17082927

[5] Sheet F. Criteria Pollutants: Particulate Matter (PM 2.5 /PM 10). 2012 :3-5

[6] Repace JL, Ott WR, Klepeis NE. Indoor air pollution from cigar smoke. Smoking and Tobacco Control Monograph No. 9. 1996;**i**(9)

[7] Krupnova TG et al. Elemental Composition of PM 2 . 5 and PM 10 and Health Risks Assessment in the Industrial Districts of Chelyabinsk. Russia: South Ural Region; 2021

[8] Warwick H, Doig A. Smoke – the Killer in the Kitchen: Indoor Air Pollution in Developing Countries. Journal of children

[9] Fowles J, Bates M, Noiton D. The Chemical Constituents in Cigarettes and Cigarette Smoke: Priorities for Harm Reduction. A report to the New Zealand Ministry of health. 2000

[10] A. Overview, Tobacco, Cigarettes and Cigarette Smoke. 2007

[11] Forbes PBC, Garland RM. Outdoor Air Pollution. Comprehensive Analytical Chemistry

[12] Admassu Mengesha, Wubeshet M. Air Pollution. British Medical Journal. 2006

[13] Winkel A et al. "Measurement of Particulate Matter : Recommendations for the VERA Test Protocol on Air Cleaning Technologies Measurement of Particulate Matter : Recommendations for the VERA Test Protocol on Air Cleaning Technologies". 2014

[14] Addisu A, Getahun T, Deti M, Negesse Y, Mekonnen B. Association of Acute Respiratory Infections with Indoor Air Pollution from Biomass Fuel Exposure among Under-Five Children in Jimma Town, Southwestern Ethiopia. Journal of Environmental and Public Health. 2021. DOI: 10.1155/2021/7112548

[15] States U, Environmental US, Agency P, The S. Carbon monoxide. Blood Coagulation and Fibrinolysis. 2000;**3**(6)

[16] Allemand N, Technical T. Guidelines for Estimation and Measurement of VOC Emissions. 2018. pp. 1-22

[17] W. H. O. E. Centre and H. Bilthoven. "Strategic Approaches to Indoor Air Policy-Making". 1998

Chapter 8

Review of Measures to Control Airborne Pollutants in Broiler Housing

José L.S. Pereira, Carla Garcia and Henrique Trindade

Abstract

Broiler housing is a significant source of airborne pollutants from animal production, which lead to degradation of indoor air quality and outdoor emissions, particularly ammonia, nitrous oxide, carbon dioxide, methane, hydrogen sulphide, odours and particulate matter. In this chapter, we first analyse the current state of the art on the consequences of these pollutants on broiler farming, farm workers, and the environment. This includes the factors affecting pollutants generation, quantification, and mitigation measures suppressing airborne pollutants. Next, we describe different best available techniques for environmental protection and sustainability of broiler production, namely feeds and feeding management, feed supplements, bedding management and treatment of exhaust air. Thus, broiler farms should select mitigation strategies based on several considerations, such as location, climate conditions, environmental policies and financial resources.

Keywords: ammonia, broiler, greenhouse gases, mitigation measure, odours, particulate matter

1. Introduction

Broiler meat is one of the main sources of protein for humans in the Mediterranean region and its consumption is expected to continue to increase until 2050. Globally, 337 million tons of meat were produced in 2019, 44% more than in 2000, with chicken meat representing more than half the increase [1]. In 2021, almost 121.5 million tons of broiler meat were produced in the World, from 74×10^9 broiler heads; in the same year, Europe produced ca. 19.5 million tons of broiler meat (11.3×10^9 heads) with Southern European countries being responsible by 17% of Europe production [2]. It has been estimated that around 10% of the carbon (C) footprint, land use and acidification potential caused by the European food basket is directly attributable to broiler meat consumption [3].

Broiler chickens have strict requirements in terms of thermal conditions and require high tech housing and management, equipped with automatic systems for feeding, drinking, heating, cooling, air conditioning and various sensors for data acquisition [4]. As can be observed in **Figure 1**, the broiler production includes farms of breeding hens for fertile eggs and of broilers for meat [5, 6]. From chicks to

IntechOpen

Figure 1.
Commercial broiler houses of breeding hens for fertile eggs (above) and broilers for meat (below).

20 weeks of age, breeding hens are housed in buildings with a solid floor and litter material (pine shavings or rice hulls). Between 20 and 60 weeks of age, hens are moved to housing with deep litter/slatted floor and the manure is removed once per year. The broilers are housed in similar buildings with a rearing cycle between 30 and 42 days (1.4–2.4 kg liveweight) [4, 7].

The broiler production is a major pollution source of reactive nitrogen (N) losses, and the emissions intensity of chicken meat is 0.6 kg CO_2eq kg^{-1} [1]. This sector is a significant source of ammonia (NH_3), nitrous oxide (N_2O) and methane (CH_4) emissions [8].

This sector is linked to NH_3, N_2O and CH_4 emissions, and have an impact on global greenhouse gas (GHG) emissions, as well as bird and human health. Litter and manure can contain pesticide residues, microorganisms, pathogens, pharmaceuticals (antibiotics), hormones, heavy-metals, macronutrients (at improper ratios) and other pollutants which can lead to air, soil and water contamination as well as formation of antimicrobial/multidrug resistant strains of pathogens. Particulate matter (PM) emitted from intensive broiler production operations contain feather and skin fragments, faeces, feed particles, microorganisms and other pollutants, which can adversely impact poultry health as well as the health of farm workers and nearby inhabitants. Odours are another problem that can have an adverse impact on health and quality of life of workers and surrounding population [9].

Farms are required to comply with environmental legislation to be managed in a sustainable and environmentally friendly way, namely: (i) the Integrated Pollution Prevention and Control (IPPC) Directive (96/61/EC), which requires large livestock production facilities to implement best available techniques (BAT) for environmental protection; (ii) the National Emissions Ceiling (NEC) Directive (2016/2284/EU), which sets emission reduction commitments relative to 2005 levels for N oxides (NOx), non-methane volatile organic compounds (NMVOC), sulphur oxides (SOx), NH_3 and $PM_{2.5}$ by 2030; (iii) the Energy and Climate Action Directive (2018/1999/EU) for carbon neutrality, which targets reducing greenhouse gas (GHG) emissions by 45–55%, based on emissions recorded in 2005 EU Commission's Joint Research Centre reference document (BREF) was introduced in 2017 and provides the BAT guidelines for intensive poultry rearing operations [10].

The aim of this review was to provide a comprehensive overview of current knowledge about the impact of airborne pollutants (NH_3, N_2O, CO_2, CH_4, hydrogen sulphide (H_2S), odours and PM) from broiler houses, including analysis and discussion of best available techniques for environmental protection and sustainability of broiler production.

2. Gases

2.1 Ammonia

Broilers excrete most of the not metabolised N in the forms of uric acid (ca. 80%), NH_3 (ca. 10%) and urea (ca. 5%). The excretions are rich in uric acid ($C_5H_4N_4O_3$), being hydrolysed into urea ($CO(NH_2)_2$) through aerobic decomposition and followed by conversion to NH_4^+ by the urease enzyme found in manure Eq. (1). Then, NH_3 is prone to be released into the air as a gas that can affect both birds and farmworkers and escape to the atmosphere by building ventilation [11, 12].

$$C_5H_4N_4O_3 \underset{Uricase}{\Rightarrow} CO(NH_2)_2 \underset{Urease}{\Rightarrow} [CO_2] + NH_4^+ \underset{Volatilisation}{\Rightarrow} NH_3 \tag{1}$$

In terms of environmental impacts, the emission of NH_3 reacts with ammonium (NH_4^+) sulphate and nitrate (NO_3^-) and chloride particles, being transported over wide areas, adversely impacting air quality and visibility [13, 14]. Also, the wet deposition of NH_4^+ salt formed after NH_3 volatilisation led to toxicity of vegetation in the surrounding farm area (<1 km), whereas the wet deposition can induce soil acidity, severely damaged biodiversity and eutrophication of water bodies [13, 14].

Ammonia volatilisation is a result of complex biological, physical and chemical processes and depends on several factors (**Figure 2**) [11]. Thus, the factors that influence how NH_3 is formed and released into the broiler house environment are litter type, animal activity, bird age and density, manure handling, frequency of manure removal, ventilation rate and air velocity, while the factors that influence how manure bacteria and enzymes break down N to form NH_3 are N content, temperature, moisture/humidity and pH (**Figure 2**) [4, 15]. The litter's pH is a major factor regulating the volatilisation of NH_3 since it specifies the volatile NH_4^+/NH_3 ratio between their ionic and non-volatile forms (**Figure 2**) [16]. Among all factors affecting NH_3 volatilisation in broiler house litter, humidity is one of the most important influencing parameters [17]. Therefore, the increase of indoor air humidity and temperature will increase the moisture and temperature of the litter material and consequently will enhance the NH_3 volatilisation [4, 18]. The water trough should be placed carefully and checked regularly to prevent water leakage on the litter and the litter humidity in broiler houses should be 15–25% [16].

Ammonia, a colourless and highly water-soluble gas, is primarily an irritant and has been known to create health problems for animals in confinement building [19]. Ammonia concentrations in broiler houses should not exceed 25 ppm because birds' productivity is adversely affected above this limit (**Table 1**). Higher NH_3 levels also decrease the body weight gain, calorie conversion, general living conditions, carcass condemnation, and birds' immune system (**Table 1**) [12, 16, 21]. Furthermore, the NH_3 gas has a common pungent odour that humans can detect at concentrations of 20–30 ppm and irritates the conjunctiva, cornea, and mucous membranes of birds'

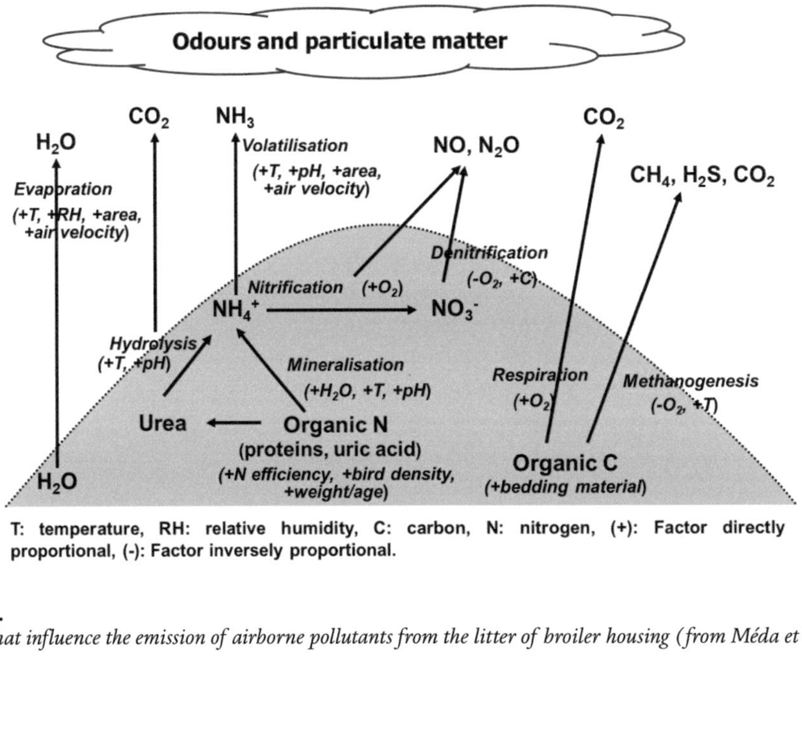

Figure 2.
Factors that influence the emission of airborne pollutants from the litter of broiler housing (from Méda et al. [15]).

Concentration	Effect
5 ppm	Lowest detectable level
6 ppm	Irritation of the eyes and the respiratory tract
11 ppm	Reduced animal performance
25 ppm	Maximum exposure level allowed for a period of one hour
35 ppm	Maximum exposure level allowed for a period of 10 minutes
40 ppm	Headache, nausea and loss of appetite in humans
50 ppm	Severe reduction in performance and animal health; increased possibility of pneumonia
100 ppm	Sneezing, salivation and irritation of mucus membranes in animals
≥300 ppm	An immediate threat to human life and health

Table 1.
Harmful effects of NH_3 concentrations in broilers and farm workers (from Birst et al. [20]).

respiratory tract at high concentrations (**Table 1**) [12, 16, 21]. High NH_3 levels can damage birds' respiratory systems' mucous membranes, increasing their susceptibility to respiratory infections, particularly to *Escherichia Coli* [12, 16, 21]. Also, high NH_3 levels increase susceptibility to Newcastle disease virus, increase the incidence of air sacculitis and keratoconjunctivitis and increase the prevalence of *Mycoplasma gallisepticum* [12, 16].

Previous studies [12, 18] recommend a limit of 10 ppm of NH_3 to maintain a good indoor air quality on broiler houses, but the threshold values of 20 ppm are recommended as limit for a short period exposure. Note that long-term NH_3 toxicity in the poultry house may increase the susceptibility of birds to the adverse effects of NH_3 even at 20 ppm (**Table 1**) [16]. However, the NH_3 gas significantly affects bird health and wellbeing, but the severity of damage depends on the concentration of NH_3 and duration of exposure (**Table 1**) [22]. Indoor NH_3 levels are also affected by housing and management factors, such as housing type, bird age and density, manure or litter conditions and handling schemes and building ventilation rate [15].

Season of the year, diurnal cycles, bird health and management practices can influence NH_3 volatilisation variability from broiler houses due to indoor and outdoor temperature and humidity differences associated to ventilation, bird activity and manure management conditions [20, 23]. Previous studies [4, 5, 7, 23, 24] observed that the lowest concentrations of NH_3 were recorded in summer period, although NH_3 emissions tended to be higher just in summer months because of a higher ventilation rate. The elevated levels of NH_3 in winter were attributed to the lower ventilation rate during cold weather. However, the average annual NH_3 emission rates from broiler houses using new litter material in each production cycle varied greatly among European countries, from 0.06 to 0.13 g day^{-1} $broiler^{-1}$ in Portugal [4, 5, 7], 0.17–0.19 g day^{-1} $broiler^{-1}$ in UK and France [25, 26] and 0.35–0.45 g day^{-1} $broiler^{-1}$ in Ireland and Spain [24, 27].

Previous studies [12, 16, 20, 28] compiled a comprehensive overview of the most promising NH_3 mitigation strategies from broiler housing such as oil and water spraying; aeration and ventilation system; filtration and biofiltration; acid scrubber; dietary manipulation; temperature and litter moisture control; immunisation; bedding materials; light intensity; manure managements; and solid additive and litter amendment. However, some of these mitigation technologies are expensive and/or needed more research to be implemented in commercial farms. Thus, **Table 2** resumes the best available techniques proposed to reduce NH_3 production in poultry houses. For Portugal, the Directive 2016/2284/EU sets a 20% NH_3 reduction from 2030 relative to 2005 levels.

2.2 Nitrous oxide

Nitrous oxide (and NO) can be released through nitrification and denitrification processes from litter Eq. (2). Nitrification is the bacterial oxidation from ammonium (NH_4^+) to NO_3^- under aerobic conditions, whereas denitrification is the reduction of NO_2^-/NO_3^- to N_2 under anaerobic conditions. The main influencing factors of nitrification and denitrification are oxygen pressure, presence off N-substrates, temperature and humidity. Consequently, litter type, stoking density and management affect the gas concentration and emission from broiler houses [15].

$$NH_4^+ \underset{Nitrification}{\Rightarrow} [N_2O]+[NO]+NO_3^- \underset{Denitrification}{\Rightarrow} [N_2O]+[NO]+N_2 \qquad (2)$$

Nitrous oxide is a colourless and non-flammable gas, with a slightly sweet odour. Known as "laughing gas" due to the euphoric effects of inhaling it, being anaesthetic and the maximum recommended indoor concentration is 3 ppm [19]. In terms of environmental impacts, N_2O in the atmosphere has a long life and contributes significantly to global warming and greenhouse effect [13, 14]. Also, contributes to the depletion of the ozone layer in the stratosphere through the photochemical decomposition of N_2O to NO [13, 14].

The rate of formation and emission of N_2O varies through time with changes in manure porosity, pH, temperature, moisture, amount of solids, N and protein content of the manure substrate (**Figure 2**) [32]. However, N_2O emission rates might be mostly related with the litter management (i.e., the interphase aerobic/anaerobic conditions of the litter), but litter temperature and protein content in the diet could enhance the N_2O loss (**Figure 2**) [4–6]. Moreover, previous studies reported low N_2O concentrations close to the detection limit levels and tended to be higher in winter than in summer, being negatively affected by the ventilation rate [4–6, 32]. The average annual N_2O emission rates from broiler houses with new litter material in each production cycle varied among European countries, ranging from 0 mg day^{-1} $broiler^{-1}$ in France [25] to 2 to 6 mg day^{-1} $broiler^{-1}$ in Portugal [4, 5] and 46 mg day^{-1} $broiler^{-1}$ in Spain [5].

There are various options to reduce N_2O emissions, but the key approach is to improve overall N efficiency of broilers production [10, 32]. Therefore, improving animal feed conversion efficiency (dry matter, fibre, protein and mineral nutrition intake) becomes a major strategy for mitigating N_2O emissions from these animal species [10, 32]. The relationship between manure NH_3 volatilisation and N_2O emission is also complex because emissions of both may be reduced by diet manipulation or manure management, and if a mitigation technology reduces NH_3 losses, the preserved NH_4^+ may later increase storage N_2O emissions [13, 33]. On the other hand, gaseous losses of N will reduce the availability of N for nitrification and denitrification processes and, consequently, N_2O formation [34]. Nevertheless, it is crucial to consider potential pollution swapping when planning and implementing mitigation measures.

Mitigation measure	Technique	NH_3	Reference
Feeding	Multiphase feeding		[13]
	Low protein feeding, with or without supplementation of specific synthetic amino acids		[13]
	Increasing the non-starch polysaccharide content of the feed		[13]
	Supplementation of pH-lowering substances, such as benzoic acid		[13]
	Decrease of the protein content in the feed by 1% may decrease the NH_3 emissions by 10%	10%	[13]
	Clinoptilolite as feed additive	30%	[29]
	Yucca extract as feed additive	50%	[29]
Housing	New and largely rebuilt broiler housing	20–90%	[13]
	Naturally ventilated house or insulated fan-ventilated house with a fully littered floor and equipped with non-leaking drinking system	20–30%	[13]
	Litter with forced manure drying using internal air	40–60%	[13]
	Tiered floor and forced air drying	90%	[13]
	Tiered removable sides; forced air drying	90%	[13]
	Combideck system	40	[13]
	Magnesium sulfate as litter additive	45%	[5]
	Aluminium sulphate as litter additive	50–70%	[13, 30]
	Clinoptilolite as litter additive	28%	[6]
	Scrubbing of exhaust air	70–90%	[13]
	Biofiltration of exhaust air	50%	[31]

Table 2.
Best available techniques (relative to the reference technique: Deep litter; fan-ventilated house) proposed to reduce NH_3 losses in broiler houses.

2.3 Carbon dioxide

Carbon dioxide emission originates from the breathing of broilers, uric acid/urea hydrolysis of excretions and aerobic/anaerobic decomposition of litter material, the first source being the main source of this emission (**Figure 2**) [4–6, 15, 18]. Carbon dioxide production by broilers is proportional to their metabolic heat production, and thus to the metabolic body weight of the broilers, which in turn is affected by the temperature and broiler activity [21, 23].

Carbon dioxide is an odourless gas and the threshold limit is set to 3000 ppm, being asphyxiant at this level, increasing breathing, drowsiness, and headaches as concentration increases [19]. Under normal conditions, the concentration of CO_2 in broiler houses ranges from 500 to 3000 ppm, with a limit of 2500 ppm of CO_2 being recommended to maintain good indoor air quality [21]. There is no health risk for birds and humans at this level. Although CO_2 is rarely life threatening, some effects are reported for longer exposure to 3000–10,000 ppm, with negative effects on blood parameters (alkaline phosphatase) and immune system in broilers and respiratory and cardiovascular diseases in humans [21]. In terms of environmental impacts, CO_2 in the atmosphere contributes significantly to global warming, but does not contribute to the greenhouse effect because it is part of the so-called short C cycle, with CO_2 originated by animal production not being accounted in national inventories [14, 35].

The CO_2 concentrations observed in broiler houses are related with NH_3, with elevated levels in winter due the lower ventilation rate during cold weather [5]. Previous studies [4, 5, 7, 21, 24] reported an average CO_2 emission rate from 55.2 to 98.4 g day^{-1} broiler^{-1} in South of Europe for broiler using new litter material in each production cycle.

2.4 Methane

Methane is produced by the anaerobic microbial degradation of soluble lipids, carbohydrates, organic acids, proteins and other organic components Eq. (3). Methane emission originates from the anaerobic decomposition of litter material inside the building, depending on temperature and humidity, ventilation rate, excretion rate and litter management (**Figure 2**) [15, 24, 32].

$$Organic\ C \underset{Anaerobic\ decomposition}{\Rightarrow} CO_2 + CH_4 + H_2S \quad (3)$$

Methane is not considered to be toxic and the threshold limit is set to 5000 ppm, being odourless and concentrations from 5000 to 15,000 ppm are explosive, with several explosions have occurred due to ignition of methane rich air in poorly ventilated animal buildings [19, 21]. In terms of environmental impacts, CH_4 in the atmosphere has a strong radiative power and contributes significantly to global warming and greenhouse effect [13, 14].

Previous studies reported low CH_4 concentrations in broiler houses and the variability may be caused by differences in litter reactions within the broiler house during the growing period [24]. The annual average CH_4 emissions ranged from 0 to 18 mg day^{-1} broiler^{-1} in South of Europe [4, 5, 24, 25] for commercial broiler houses using new litter material in each production cycle, showing lower values than the IPCC emission factor (50 mg day^{-1} broiler^{-1}) [14].

2.5 Hydrogen sulphide

Hydrogen sulphide is formed by bacterial sulphate reduction and the decomposition of sulphur containing organic compounds in litter under anaerobic conditions (**Figure 2**) [36].

The H_2S as low odour thresholds (10 ppb) and when managed improperly, higher concentrations negatively affect humans, birds, and the environment. At low concentrations (<10 ppm), this gas is highly toxic, poisonous, deadly, odorous (odour of rotten eggs/low concentrations contributed significantly to odour), colourless, and heavier than air. The H_2S could cause dizziness, headaches, and irritation to the eyes and the respiratory tract. In addition to causing adverse effects to human and bird health, H_2S might be oxidised in the air forming sulphuric acid (H_2SO_4) resulting in acid rain that could cause ecological damage [19, 21].

Saksrithai and King [36] summarised a comprehensive overview of the most promising H_2S mitigation strategies from broiler, including feed supplementation (additives, prebiotics, and probiotics); manure manipulation (pH, moisture, and its microbial population); housing types; ventilation rates; and biofilters. In addition, the most promising singular methods to reduce 100% H_2S emissions are probiotic supplementation in feed, sawdust in manure, or a biofiltration system. Where cost and equipment availability may be prohibitive, combined methods (assuming additive effects) of fibrous by-products and manure moisture control via microorganisms or oil addition can reduce 100% emissions as well.

3. Odours

The odours are a significant source of gaseous pollution in broiler housing, derived from anaerobic and aerobic microbial activity during litter decomposition. The litter is considered the primary source of odour in broiler housing. The malodorous compounds are originated by organic particulate matter, volatile fatty acids, sulphurous (H_2S, mercaptans) and NH_3 (**Figure 2**) [37]. The main odorous compounds emitted on broiler farms are dimethyl sulphide, dimethyl disulphide, dimethyl trisulfide, nhexane, acetic acid, 2,3-butanedione, methanol, ethanol, 1-butanol, 2- butanol, 1-octen-3-ol, 3-methyl-1-butanol, 3-methyl-1-butanal, acetone, 2- butanone and 3-hydroxy-2-butanone [38].

Odour emissions from litter are complex due to the existence of multiple odorant sources within litter (i.e., fresh excreta, friable litter and cake), formation and emission of numerous odorants, and significant spatial and temporal variability of moisture content, porosity, pH, ventilation airflow, temperature, humidity and bird activity (**Figure 2**) [9, 38].

There has been limited studies of management strategies that reduce the formation and emission of odorants from broiler litter, mostly focussed on the perspective of reducing NH_3 emissions by air scrubbing, misting, filtering, ionising, oxidising and dispersing technologies [28, 38, 39]. In addition, strategies are focused on the perspective of reducing odours in the air through fogging technologies combined with the use of masking agents, counteractants, neutralizers and surface-enhanced absorption agents (**Table 3**) [28, 38, 39]. Dunlop et al. [40] selected management strategies with expected effectiveness to reduce odour emissions from broiler litter, being maintaining dry and friable litter; *in situ* aeration of litter; in shed windrowing/pasteurising of litter reuse; litter acidifying additives; litter adsorbent addition (activated carbon, silica gel or zeolite); and enzyme addition combined with heated incubation.

4. Particulate matter

In broiler housing, the PM consists of a complex mixture of solid and liquid materials such as litter materials, feathers, feeds, skin, excreta, dander and microorganism, whit about 90% organic content [41]. The classification of PM is based on particle size (aerodynamic diameter), considering $PM_{1.0}$ ($\leq$1.0 μm), $PM_{2.5}$ ($\leq$2.5 μm), $PM_{4.0}$ ($\leq$4.0 μm), $PM_{10.0}$ ($\leq$10.0 μm) and total suspended particle (TSP) ($\leq$100.0 μm) [42]. The PM emissions from broiler housing are affected by various factors and change according to housing system or types, litter materials, diurnal and seasonal variation, ventilation system and velocity, temperature and relative humidity, birds age and type, activity and stocking density and manure management (**Figure 2**) [41].

The primary air emissions include PM with a high potential risk to air quality, public and bird health, and climate change. Cambra-López et al. [43] reported that the concentrations of TSP (inhalable PM) in broiler houses ranged from 1 to 14 mg m^{-3} in European countries. The World Health Organisation has recently (in 2021) amended the ambient air quality standards and proposed the maximum of $PM_{10.0}$ to be 15 µg m^{-3} for the annual average and 45 µg m^{-3} for the 24 h mean, while for $PM_{2.5}$ to be 5 µg m^{-3} for the annual average and 15 µg m^{-3} for the 24 h mean [41]. In Portugal, the Directive 2016/2284/EU establishes a 53% reduction of $PM_{2.5}$ from 2030 compared to 2005 levels. The levels of PM in broiler houses affect birds' health and welfare, including eye irritation, throat irritation, cough, phlegm, chest tightness, sneezing, headache, fever, nasal congestion, and wheezing, especially in cold periods when the house will have limited ventilation. Furthermore, long term exposure to PM in humans increases obstructive pulmonary disorder, chronic bronchitis, chronic obstructive pulmonary disease, pneumonia lesions, cardiovascular disease, asthma like symptoms, lung cancer, or even mortality. Similarly, a higher level of PM with endotoxin in birds causes impaired lung function, chronic bronchitis, pneumonia lesions, cardiovascular illness, and cardiotoxicity in chicken embryos and hatchling chickens and might increase the risk of mortality rates [21, 41, 43].

Principle	Technique	Pollutant
Microbiological	Aeration	Gases, odour
	Anaerobic digestion	Gases, odour
	Biofiltration	Gases, odour, PM
	Composting	Gases, odour
Biochemical	Enzyme additives	Gases, odour
Chemical	Acidification	NH_3
	Ozonation	Odour
Managerial	Best management practices	Gases, odour, PM
Physical	Absorption	Gases, odour
	Cooling	NH_3
	Covering	Gases, odour
	Drying	Gases, odour, PM
	Electrostatic precipitation	PM
	Fan plum deflector	Gases, odour, PM
	Filtering	PM
	Oil spraying	Gases, odour, PM
	Shelterbelt	Gases, odour, PM
	Solid separation	Gases, odour, PM
	Wet scrubbing	Gases, odour, PM
Physiological	Dietary manipulation	NH_3, odour
	Odour masking	Odour

Table 3.
Best available technologies for mitigation of airborne pollutants in broiler houses (from Ni [28]).

Table 3 show effective mitigation strategies to reduce PM in broiler houses using biochemical, chemical, managerial, physical, and physiological practices. The techniques to control PM will also reduce NH_3 and odours, which can be managing housing system and cleaning, light intensity, oil and water spraying, litter materials, electrostatic ionisation, filtration and biofiltration, acid scrubber, windbreaks and vegetative shelterbelts [28, 41, 43, 44].

5. Conclusions

In this review study, were summarised the primary reasons that affect the emission of NH_3, N_2O, CO_2, CH_4, H_2S, odours and PM in housing, then analysed the consequences of these pollutants on birds, workers and environment, and finally discussed different best available techniques for environmental protection and sustainability of broiler production, particularly feeds and feeding management, feed supplements, litter management and treatment of exhaust air. Thus, broiler farms should select mitigation strategies based on several considerations, such as location, climate conditions, environmental policies and financial resources.

Acknowledgements

This work is supported by National Funds by FCT - Portuguese Foundation for Science and Technology, under the project UIDB/04033/2020.

Conflict of interest

The authors declare no conflict of interest.

Author details

José L.S. Pereira[1,2]*, Carla Garcia[1] and Henrique Trindade[2]

1 Agrarian School of Viseu, Polytechnic Institute of Viseu, Quinta da Alagoa, Viseu, Portugal

2 Centre for the Research and Technology of Agro-Environmental and Biological Sciences (CITAB), Inov4Agro, University of Trás-os-Montes and Alto Douro, Quinta de Prados, Vila Real, Portugal

*Address all correspondence to: jlpereira@esav.ipv.pt

IntechOpen

© 2023 The Author(s). Licensee IntechOpen. This chapter is distributed under the terms of the Creative Commons Attribution License (http://creativecommons.org/licenses/by/3.0), which permits unrestricted use, distribution, and reproduction in any medium, provided the original work is properly cited.

References

[1] Food and Agriculture Organization of the United Nations (FAO). World Food and Agriculture - Statistical Yearbook 2021. Rome: FAO; 2021. p. 368

[2] Instituto Nacional de Estatística (INE). Estatísticas Agrícolas 2020. Lisboa, Portugal: Instituto Nacional de Estatística, I.P; 2020. p. 181

[3] Notarnicola B, Tassieli G, Renzulli PA, Castellani V, Sala S. Environmental impacts of food consumption in Europe. Journal of Cleaner Production. 2017;**140**:753-765

[4] Pereira JLS, Ferreira S, Pinheiro V, Trindade H. Ammonia, nitrous oxide, carbon dioxide and methane emissions from commercial broiler houses in Mediterranean Portugal. Water, Air & Soil Pollution. 2018;**229**:377

[5] Pereira JLS, Ferreira S, Pinheiro V, Trindade H. Ammonia and greenhouse gas emissions following the application of clinoptilolite on the litter of a breeding hen house. Environmental Science and Pollution Research. 2019;**26**:8352-8357

[6] Pereira JLS, Ferreira S, Pinheiro V, Trindade H. Effect of magnesium sulphate addition to broiler litter on the ammonia, nitrous oxide, carbon dioxide and methane emissions from housing. Atmospheric Pollution Research. 2019;**10**:1284-1290

[7] Pereira JLS. Assessment of ammonia and greenhouse gas emissions from broiler houses in Portugal. Atmospheric Pollution Research. 2017;**8**:949-955

[8] Portuguese Informative Inventory Report (PIIR). Portuguese Informative Inventory Report 1990-2020. Submitted under the NEC Directive (EU) 2016/2284 and the UNECE Convention on Long-range Transboundary Air Pollution. Portuguese Informative Inventory Report (PIIR). Portuguese Environmental Agency, Lisbon, Portugal. 2022. p. 683

[9] Gržinić G, Piotrowicz-Cieślak A, Klimkowicz-Pawlas A, Górny RL, Ławniczek-Wałczyk A, Piechowicz L, et al. Intensive poultry farming: A review of the impact on the environment and human health. Science of The Total Environment. 2023;**858**:160014

[10] European Commission (EC). Best Available Techniques (BAT) Reference Document for the Intensive Rearing of Poultry or Pigs: Industrial Emissions Directive 2010/75/EU (Integrated Pollution Prevention and Control). JRC107189. 2017. p. 898

[11] Hartung J, Phillips VR. Control of gaseous emissions from livestock buildings and manure stores. Journal Agricultural Engineering Research. 1994;**57**:173-189

[12] Naseem S, King AJ. Ammonia production in poultry houses can affect health of humans, birds, and the environment-techniques for its reduction during poultry production. Environmental Science and Pollution Research. 2018;**25**:15269-15293

[13] Bittman S, Dedina M, Howard CM, Oenema O, Sutton MA. Options for Ammonia Mitigation: Guidance from the UNECE Task Force on Reactive Nitrogen; Project Reference: CEH Project no. C04910. Edinburgh, UK: NERC/Centre for Ecology & Hydrology; 2014. p. 83

[14] The Intergovernmental Panel on Climate Change (IPCC). In: Calvo Buendia E, Tanabe K, Kranjc A,

Baasansuren J, Fukuda M, Ngarize S, Osako A, Pyrozhenko Y, Shermanau P, Federici S, editors. Refinement to the 2006 IPCC Guidelines for National Greenhouse Gas Inventories. Geneva, Switzerland: IPCC; 2019

[15] Méda B, Hassouna M, Aubert C, Robin P, Dourmad JY. Influence of rearing conditions and manure management practices on ammonia and greenhouse gas emissions from poultry houses. World's Poultry Science Journal. 2011;**67**:441-456

[16] Swelum AA, El-Saadony MT, Abd El-Hack ME, Ghanima MMA, Shukry M, Alhotan RA, et al. Ammonia emissions in poultry houses and microbial nitrification as a promising reduction strategy. Science of The Total Environment. 2021;**781**:146978

[17] Carey J, Lacey R, Mukhtar S. A review of literature concerning odors, ammonia, and dust from broiler production facilities: 2. Flock and house management factors. Journal of Applied Poultry Research. 2004;**13**:509-513

[18] Pereira JLS, Ferreira S, Garcia CSP, Conde A, Ferreira P, Pinheiro V, et al. Assessment of ammonia and carbon dioxide concentrations in a breeding hen building under Portuguese winter. International Journal of Food and Biosystems Engineering. 2017;**5**:1-6

[19] Samer M. Abatement techniques for reducing emissions from livestock buildings. Springer Briefs in Environmental Science. 2016;**2016**:62

[20] Birst RB, Subedi S, Chai L, Yang X. Ammonia emissions, impacts, and mitigation strategies for poultry production: A critical review. Journal of Environmental Management. 2023;**328**:116919

[21] International Commission of Agricultural Engineering (CIGR). Aerial Environment in Animal Housing. Working Group Report Series No 94.1. Rennes, France: International Commission of Agricultural Engineering (CIGR) and CEMAGREF; 1994. p. 116

[22] Miles DM, Branton SL, Lott BD. Atmospheric ammonia is detrimental to the performance of modern commercial broilers. Poultry Science. 2004;**83**:1650-1654

[23] Brouček J, Čermák B. Emission of harmful gases from poultry farms and possibilities of their reduction. Ekológia. 2015;**3**:89-100

[24] Calvet S, Cambra-López M, Estellés F, Torres AG. Characterization of gas emissions from a Mediterranean broiler farm. Poultry Science. 2011;**90**:534-542

[25] Guiziou F, Béline F. In situ measurement of ammonia and greenhouse gas emissions from broiler houses in France. Bioresource Technology. 2005;**96**:203-207

[26] Misselbrook TH, Gilhespy SL, Cardenas LM, Williams J. Dragosits U. Inventory of ammonia emissions from UK agriculture. DEFRA. 2015;**2016**:36

[27] Hayes ET, Curran TP, Dodd VA. Odour and ammonia emissions from intensive poultry units in Ireland. Bioresource Technology. 2006;**97**:933-939

[28] Ni JQ. Research and demonstration to improve air quality for the U.S. animal feeding operations in the 21st century — A critical review. Environmental Pollution. 2015;**200**:105-119

[29] Amon M, Dobeic M, Sneath RW, Phillips VR, Misselbrook TH, Pain BF. A farm-scale study on the use of clinoptilolite zeolite and De-Odorases for

reducing odour and ammonia emissions from broiler houses. Bioresource Technology. 1997;**61**:229-237

[30] Eugene B, Moore PA, Li H, Miles D, Trabue S, Burns R, et al. 2015. Effect of alum additions to poultry litter on in-house ammonia and greenhouse gas concentrations and emissions. Journal of Environmental Quality. 2005;**44**:1530-1540

[31] Van der Heyden C, Demeyer C, Volcke EIP. Mitigating emissions from pig and poultry housing facilities through air scrubbers and biofilters: State-of-the-art and perspectives. Biosystems Engineering. 2005;**134**:74-93

[32] Brouček J. 2018. Nitrous oxide release from poultry and pig housing. Polish Journal of Environmental Studies. 2018;**27**:467-479

[33] Petersen SO, Sommer SG. Ammonia and nitrous oxide interactions: Roles of manure organic matter management. Animal Feed Science and Technology. 2011;**166-167**:503-513

[34] Hristov AN, Oh J, Lee C, Meinen R, Montes F, Ott T, et al. In: Gerber PJ, Henderson B, Makkar HPS, editors. FAO Animal Production and Health Paper No. 177Mitigation of Greenhouse Gas Emissions in Livestock Production – A Review of Technical Options for Non-CO_2 Emissions. Rome, Italy: FAO; 2013. p. 232

[35] Pereira J, Fangueiro D, Misselbrook TH, Chadwick DR, Coutinho J, Trindade H. Ammonia and greenhouse gas emissions from slatted and solid floors in dairy cattle houses: A scale model study. Biosystems Engineering. 2011;**109**:148-157

[36] Saksrithai K, King AJ. Controlling hydrogen Sulfide emissions during poultry productions. Journal of Animal Research and Nutrition. 2018;**3**:2

[37] Ni JQ, Erasmus MA, Croney CC, Li C, Li Y. A critical review of advancement in scientific research on food animal welfare-related air pollution. Journal of Hazardous Materials. 2021;**408**:124468

[38] Dunlop MW, Blackall PJ, Stuetz RM. Odour emissions from poultry litter – A review litter properties, odour formation and odorant emissions from porous materials. Journal of Environmental Management. 2016;**177**:306-319

[39] Bouzalakos S, Jefferson B, Longhurst PJ, Stuetz RM. Developing methods to evaluate odour control products. Water Science and Technology. 2004;**50**:225-232

[40] Ullman JL, Mukhtar S, Lacey RE, Carey JB. A review of literature concerning odors, ammonia, and dust from broiler production facilities: 4. Remedial management practices. Journal of Applied Poultry Research. 2004;**13**:521-531

[41] Birst RB, Chai L. Advanced strategies for mitigating particulate matter generations in poultry houses. Applied Sciences. 2022;**12**:11323

[42] Chai L, Xin H, Wang Y, Oliveira J, Wang K, Zhao Y. Mitigating particulate matter generation in a commercial cage-free hen house. Transactions of the ASABE. 2019;**62**:877-886

[43] Cambra-López M, Aarnink AJ, Zhao Y, Calvet S, Torres AG. Airborne particulate matter from livestock production systems: A review of an air pollution problem. Environmental Pollution. 2010;**158**:1-17

[44] Patterson PH. Management strategies to reduce air emissions: emphasis – dust and ammonia. Journal of Applied Poultry Research. 2005;**14**:638-650

Chapter 9

Air Quality in Mexico City after Mayor Public Policy Intervention

Jorge Méndez-Astudillo, Ernesto Caetano and Karla Pereyra-Castro

Abstract

Air pollution can be produced from anthropogenic or natural sources. Most of the policies enacted to improve air quality focus on reducing anthropogenic sources of pollution, but if natural sources increase, then air quality does not improve with these policies. In this chapter, we first define the diurnal and monthly cycle of particulate matter and ozone concentration, depending on the weather, using data from air quality monitoring stations from Greater Mexico City. We then look at a mayor public policy intervention during the COVID-19 pandemic that drastically reduced anthropogenic sources of PM but did not reduce natural sources by doing robust trend analysis on air quality station data. We evaluate the effect of these interventions by looking at national air quality standards and the number of days air pollutants have been within recommended levels. The results show that during lockdown, air quality improved because less anthropogenic sources of PM were active. However, natural sources contributed to air pollution during that time.

Keywords: particulate matter, ozone, policy intervention, Mann-Kendall trend analysis, air quality

1. Introduction

Greater Mexico City (GMC) is one of the largest cities in the world, with more than 22 million inhabitants [1]. The population density in the capital of the country is 6163.3 inhabitants per square kilometer. Municipalities located to the east and north of GMC (e.g., Iztapalapa and Gustavo A. Madero) are more populated than those in the south (e.g., Milpa Alta). The combined activity of vehicles and industries consumes more than 45 million liters of petroleum fuel per day, generating thousands of tons of pollutants.

The labor dynamics of the Mexico City metropolitan area include long daily commutes in search of better salaries and working conditions. The cause of the mobility toward the central metropolis is explained by the diversity and specialization of the labor market, as well as the better salaries offered there. Mobility requires the use of public or private transport, whose emissions are difficult to control. Emissions from industry or vehicular traffic, when combined with the climatic and topographic characteristics of the region, lead to the production of pollutants [2]. Ozone (O_3) and particulate matter (PM_{10} and $PM_{2.5}$) are two criteria pollutants whose concentrations have remained elevated over time, creating environmental contingencies in GMC.

The formation of O_3 from pollutants has been studied, based on physics, chemistry, and statistics [3–5]. Ozone is formed by the photochemical reaction of volatile organic compounds (VOCs) and nitrogen oxides (NOx). Previous studies have reported that the formation of O_3 is sensitive to VOCs in the GMC, explaining the decrease in the rate of titration of O_3 by NO and the decrease in NOx in a VOC-limited environment. Control strategies to reduce VOC emissions will decrease ozone concentrations in VOC-limited regimes but increase their formation and concentration in NOx-limited areas [6].

Velasco et al. [7] found that in GMC, the release of heat stored in the urban surface forms a shallow stable layer (~200 m) near the ground, which favors the stagnation of nocturnal emissions. Strong inversion layers occur in the atmosphere of GMC during the night and early morning hours. After sunrise, surface heating favors convection and layer mixing, then the O_3 balance depends on the photochemical production of the pollutant, the entrainment from the upper residual layer, and the destruction by titration with nitric oxide.

No reduction in ozone concentrations above GMC is observed on weekends when the number of cars on the roads is lower than during the week. Enforcement of traffic rules that restrict car circulation (with the goal of NO_x reduction) during environmental contingencies does not necessarily reduce ozone production [8]. Improving air quality in the GMC requires the implementation of comprehensive measurements at the regional scale.

Lei et al. [5] explain that the O_3 formation characteristics and sensitivity to emission changes were found to be weakly dependent on the meteorological conditions for GMC. The O_3 formation is sensitive to NO_x and VOC levels and to the photochemical plume transport pathway.

During the COVID-19 pandemic shutdown, emissions of primary criteria pollutants at GMC were significantly reduced; however, the daily mean ozone concentration profile was similar to that of previous non-pandemic years. The reductions in NOx were so drastic that ozone formation quickly shifted from a VOC-sensitive regime to a NOx-sensitive regime. A VOC-sensitive regime means that an increase in VOC leads to an increase in O_3, while an increase in NO_x leads to a decrease of O_3; this regime is typical of densely populated urban atmospheres such as GMC [9].

Dust particles contribute to PM_{10} concentrations in GMC, particularly in the northeastern part of the city, where geologic material from the dry Lake of Texcoco is a dust source [10, 11]. This source alone generated about 80% of the total coarse particles measured in the northeastern GMC during exceptional dust events [12]. Another source of dust is the agricultural areas of Tenango del Aire and Chalco. These regions affected the central, southern, and southeastern parts of the city and contributed about 75% of the total coarse particles [12]. $PM_{2.5}$ and PM_{10} affect the air quality during the rainy season when rain removes pollutants from the air. Bare soil and traffic-related conditions from February to May also contribute to the increasing concentration of $PM_{2.5}$ and PM_{10} [13].

In this chapter, we evaluate the impact of policy interventions on anthropogenic pollution sources in the air quality of Mexico City. First, ground-based particulate matter (PM_{10}, $PM_{2.5}$) and ozone (O_3) concentration data are used to define the diurnal and annual cycle of air pollution concentrations in Mexico City. Then, the effect of the mobility restrictions enacted in March 2020 to stop the spread of the COVID-19 pandemic is evaluated through trend analysis of the time series of the pollutants from 2012 to 2022. Finally, the number of days exceeding the National Air Quality Standard is presented to evaluate the effect of reducing anthropogenic sources of pollution during March-April 2020.

2. Study area and data

2.1 Study area

According to the latest census in 2020, Mexico City (19.43°N, 99.13°W) is home to 9,210,000 inhabitants [1]. Greater Mexico City (GMC) includes some municipalities of the State of Mexico that are attached to Mexico City, with a total population of about 22 million inhabitants in 2020 [1]. In terms of weather, the GMC experiences middle-latitude systems during the dry season and is exposed to tropical interactions during the rainy season. The wet season lasts from June to November and the dry season is defined from December to May [13].

In terms of urbanization, the layout of the GMC is heterogeneous, meaning that areas depending on land use or land cover are not well defined and they are rather mixed. In this study, we defined urban areas as areas with many built structures such as buildings and roads, rural areas are those with a few built structures and mostly vegetation cover, and semi-urban areas are defined as large green areas within an urban area, such as a large park near the city center [13].

In terms of mobility, in the morning at around 08:00 h local time (LT), most people travel from the municipalities further away from the city to the city center and the eastern business district. Moreover, from 17:00 h LT the flow population is in the opposite direction, from downtown and business district to the outskirts of Mexico City, as reported in the official 2019 mobility report [14].

Figure 1 shows a satellite image of the GMC and the location of the air quality monitoring stations used in this study.

2.2 Ground-based data

The Government of Mexico City, through the Secretary of the Environment, operates the Meteorology and Solar Radiation Monitoring Network (REDMET) [15] and the Automatic Ambient Monitoring Network (RAMA) [16], which are used to monitor the weather and air pollution concentrations, respectively.

2.2.1 Air temperature data

The REDMET measures surface temperature (2 m), relative humidity (2 m), wind direction, and wind speed (10 m) every hour at 26 locations in GMC [15]. REDMET has been in operation since 1986; in 2015, some stations were decommissioned, while new stations were commissioned to cover the entire metropolitan area. To evaluate the effect of meteorological conditions on PM concentrations, 5 REDMET weather stations that report meteorological and pollution concentration data at the same location were selected. The study period was 2012–2022. In some cases, the selected stations report only one type of pollutant, such as PM_{10}, $PM_{2.5}$, or O_3.

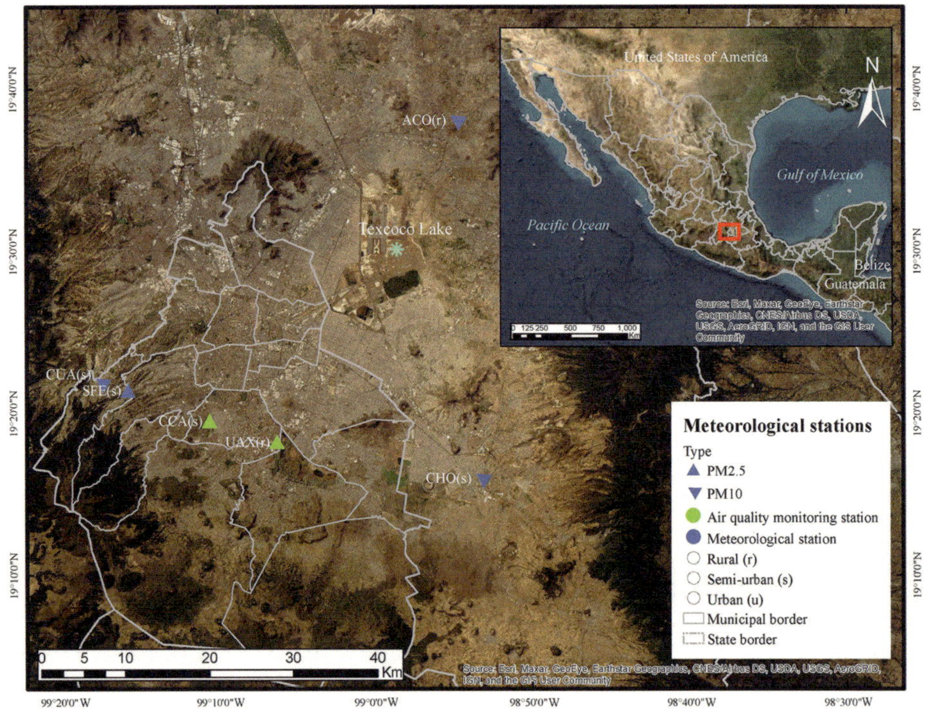

Figure 1.
Location of stations used in this study.

2.2.2 PM and O_3 concentration data

The RAMA network includes 44 air quality monitoring stations covering the entire Mexico City metropolitan area. It reports hourly concentrations of O_3, NO_2, NOx, NO, SO_2, CO, PM_{10}, and $PM_{2.5}$ [16]. We selected eight stations (**Figure 1**) with data for the period 2012–2022. Two stations (MER and HGM) are in urban areas and report PM_{10}, $PM_{2.5}$, and O_3 concentrations. We selected two stations located in rural areas (ACO and UAX), which report PM_{10} and O_3 and $PM_{2.5}$ and O_3 concentrations, respectively. Stations CHO, CUA, CCA, and SFE are in semi-urban areas, and the first two report PM_{10} and O_3 concentrations, while the other two reports $PM_{2.5}$ and O_3 concentrations.

2.3 Methodology

Hourly and monthly averages of PM (10 and 2.5) and ozone concentrations in urban, rural, and semi-urban areas were obtained to define the diurnal and monthly cycles of these pollutants. Similarly, hourly and monthly averages of meteorological data from REDMET were obtained to study the effect of meteorological conditions on PM and ozone concentrations.

The effect of mobility restrictions in 2020 on PM and O_3 concentrations, which can be considered as a major public policy intervention, was evaluated using the Mann-Kendall trend analysis with Sen's slope [17]. In addition, the Chow [18] test, which is used to test whether a break occurs at a given time in a time series, was used to assess the effect of the 2020 mobility restrictions. Statistical significance of PM concentration differences was assessed using the nonparametric Mann-Whitney U test, which is commonly used in air quality studies to compare differences in pollutant levels [19, 20]. A 95% significance level ($p < 0.05$) was used for all tests. Finally, the number of days exceeding the national air quality standard is used to determine the effect of public policy intervention on PM concentrations in GMC.

3. Results

3.1 Effect of meteorological variables on particulate matter and ozone concentrations

Multiple linear regression between meteorological variables (temperature, relative humidity RH, wind speed, and wind direction) and PM (10 and 2.5) and ozone was performed to assess the effect of meteorological variables on PM and O_3 concentrations. In 90% of the linear regression models, wind speed and wind direction were statistically significant variables, whereas temperature and RH were not always statistically significant. The effect of meteorological variables on PM and on ozone concentrations was analyzed using the quantile regression at 75% between wind speed and pollutant concentration [21].

Table 1 indicates that there is a negative relation between PM_{10} in urban areas and wind speed. Similarly, a negative relation was found between $PM_{2.5}$ in urban areas and wind speed. In semi-urban areas, a positive relation between $PM_{2.5}$ and wind speed was found. Moreover, in rural and semi-urban areas, O_3 was positively related to wind speed. Higher wind speeds in urban areas cause a higher dilution of pollutants in the Planetary Boundary Layer (PBL) decreasing PM concentrations. However, higher winds in rural and semi-urban areas enhance soil erosion, producing dust and increasing PM concentrations.

	Station	Coefficient	p-value
PM_{10}	Urban	Neg	$p < 2e-6$
	Rural	NS	$p = 0.861$
	Semi-urban	NS	$p = 0.412$
$PM_{2.5}$	Urban	Neg	$p < 2e-6$
	Rural	NS	$p = 0.426$
	Semi-urban	Pos	$p = 0.01$
O_3	Urban	NS	$p = 0.91$
	Rural	Pos	$p < 2e-6$
	Semi-urban	Pos	$p < 2e-6$

Neg: negative coefficient, Pos: positive coefficient, NS: not statistically significant ($p < 0.05$).

Table 1.
Results of quartile regression between pollutants (PM_{10}, $PM_{2.5}$, and O_3) and wind speed.

3.2 Pollutant concentration patterns

3.2.1 Hourly averages

Hourly mean pollution concentrations for stations in urban, semi-urban, and rural stations are shown in **Figure 2** for June to January and **Figure 3** for February to May.

During the first hours of the day (0:00 to 7:00 LT), the median PM_{10} concentration in the semi-urban area is lower than that recorded in the rural and urban areas. Between 9:00 h and 15:00 LT, the median PM_{10} concentration in the urban area increases and exceeds that of PM_{10} in other regions. At night (18:00–22:00 h LT), the PM_{10} concentration in the city becomes homogenous. This variability occurs during the dry and wet seasons and is more pronounced during the dry season (**Figure 2**). The hourly behavior of $PM_{2.5}$ is similar to that of PM_{10}. However, the difference in $PM_{2.5}$ concentration between the urban area and the semi-urban and rural areas is more pronounced from 9:00 to 12:00 LT (**Figure 3**).

The diurnal variation of ozone is mainly due to photochemical reactions due to solar radiation. Therefore, the ozone concentration increases from 9:00 local time to a maximum of around 15:00 local time. At night (20:00 to 7:00 LT), the ozone concentration is the lowest in the urban area compared to other regions (**Figures 2** and **3**).

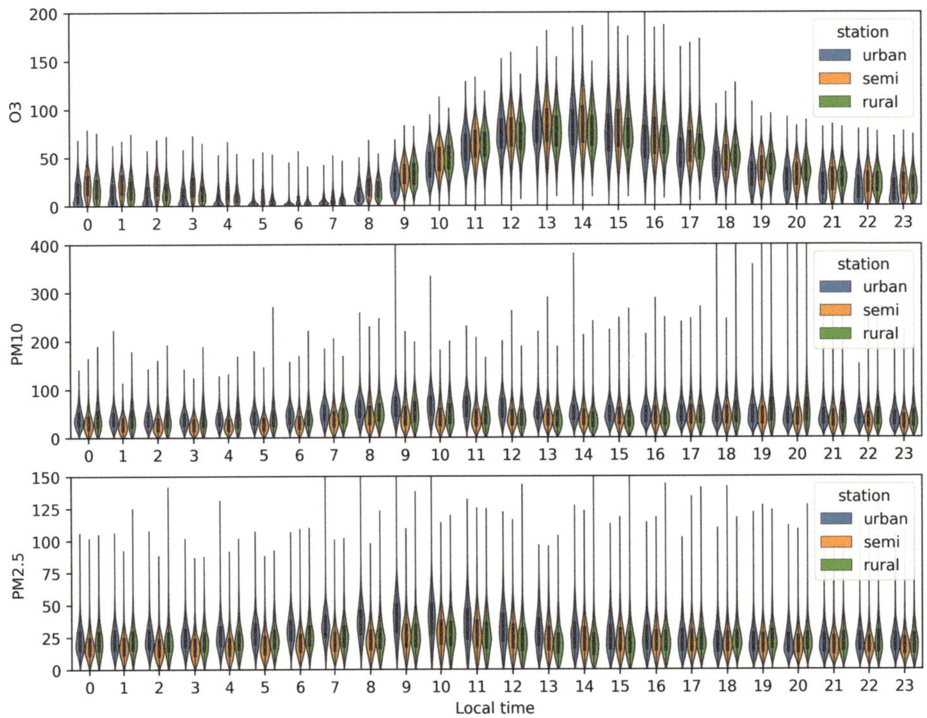

Figure 2.
Diurnal cycle of PM_{10}, $PM_{2.5}$, and O_3 concentrations in urban, semi-urban, and rural areas in GMC during the dry season (February-May). Data from 2012 to 2022. O_3 concentrations in ppb and PM (10 and 2.5) in µg/m³.

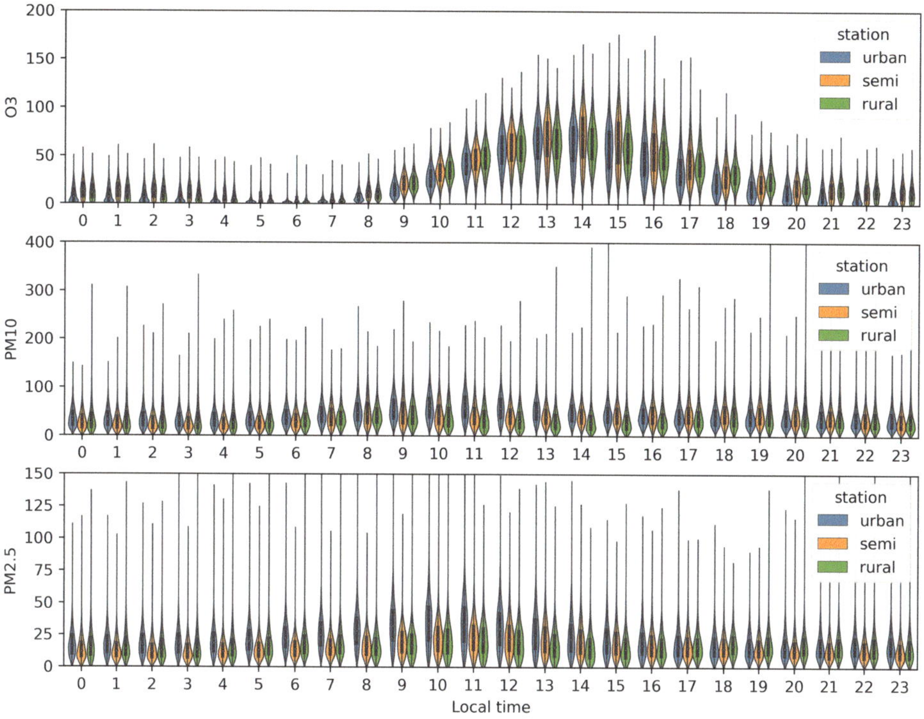

Figure 3.
Diurnal cycle of PM_{10}, $PM_{2.5}$, and O_3 concentrations in urban, semi-urban, and rural areas in GMC during the wet season (June-January) for the period 2012–2022. O_3 concentrations in ppb and PM (10 and 2.5) in μg/m³.

3.2.2 Monthly averages of pollution concentrations in GMC

Figure 4 shows the monthly average concentrations of pollutants in urban, semi-urban, and rural areas of the GMC.

Monthly ozone concentrations increased during the dry season (February-May) in urban, semi-urban, and rural areas during 2012–2022. The rainy season begins in late May in central and southern Mexico. The rain removes pollutants from the atmosphere, resulting in a decrease in tropospheric ozone concentration beginning in June. Monthly ozone concentrations are higher in rural and semi-urban areas than in urban areas (**Figure 4**). This urban-rural gradient has also been observed in other cities around the world and originates from the ratio of NOx to VOCs [22, 23].

Similarly, PM_{10} concentrations increase during the dry season. However, during this period, PM_{10} concentrations are higher in semi-urban areas than in rural areas. Furthermore, during winter (October-January), PM_{10} concentrations do not differ significantly between rural, urban, and semi-urban areas. The seasonal behavior of $PM_{2.5}$ is similar to that of PM_{10}. The median $PM_{2.5}$ concentration is 25 μg/m³ from November to May.

3.3 Trend analysis

The trend analysis shows that O_3 in urban areas in the wet season has no trend change in the period 2012–2022. Also, $PM_{2.5}$ concentrations in the dry season in rural

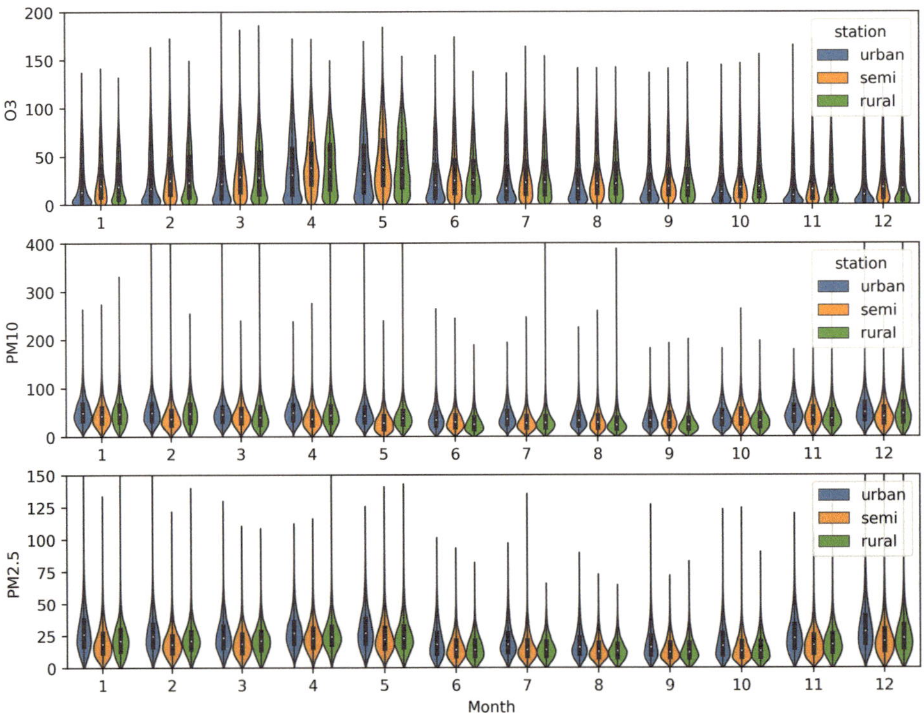

Figure 4.
Monthly ozone (O_3), PM_{10}, and $PM_{2.5}$ concentrations in urban, semi-urban, and rural areas of Mexico City for the period 2012—2022. O_3 concentrations in ppb and PM (10 and 2.5) in µg/m³.

and semi-urban areas have the same trend in the period 2012–2022. In contrast, $PM_{2.5}$ concentrations in urban areas in dry and wet seasons changed six and four times in the period 2012–2022, respectively. O_3 concentrations in rural areas in dry and wet seasons changed five times in the period 2012–2022.

In 2020, there are significant ($p < 0.05$) trend changes in $PM_{2.5}$ concentrations in urban areas during the dry season and O_3 concentrations in rural and semi-urban areas during the wet season.

3.4 Concentration difference

According to the boxplots shown in **Figure 5**, the concentration of PM_{10} is decreasing in urban, semi-urban, and rural areas compared to previous years. However, the largest decrease was found in rural areas in 2020. A decrease of 38.87% ($U = 17{,}156$; $p < 5.02e\text{-}68$) was found in 2020 compared to 2019. $PM_{2.5}$ concentrations in urban areas decreased on average by 9.7% compared to 2019. $PM_{2.5}$ in rural and semi-urban areas decreased on average by 11.68% and 14.84%, respectively ($U = 17{,}156$; $p > 5.02e\text{-}68$ and $U = 58{,}648$; $p = 0.002$). In contrast, ozone concentrations increased on average by 3% in urban, 14% in rural, and 16% in semi-urban stations in 2020 compared to 2019, due to the chemical reaction explained in detail by Peralta et al. [9].

Average increases or changes of PM (10 and 2.5) and O_3 were calculated using data from the whole year 2020. In order to assess the immediate effect of a major intervention on PM and ozone concentrations in urban, rural, and semi-urban areas of the

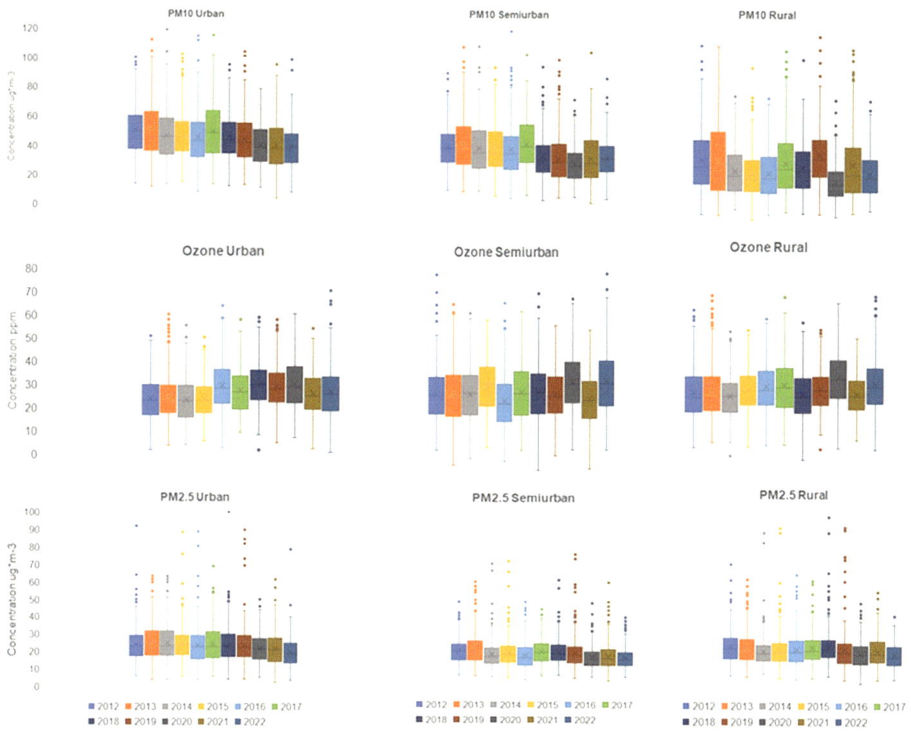

Figure 5.
Daily PM_{10} and $PM_{2.5}$ concentrations and hourly O_3 concentrations in urban, rural, and semi-urban areas of Mexico City.

At day of the year (DOY) 83, March 23rd, 2020			
	PM_{10}	$PM_{2.5}$	O_3
Urban	F = 2.5954, p-value = 0.076	F = 1.3796, p-value = 0.253	F = 151.42, p-value <2.2e-16
Rural	F = 0.7036, p-value = 0.4955	F = 2.3805, p-value = 0.09396	F = 93.78, p-value <2.2e-16
Semi-urban	F = 1.7961, p-value = 0.1675	F = 4.3136, p-value = 0.01407	F = 115.11, p-value <2.2e-16

Table 2.
Results of Chow test ($p < 0.05$).

GMC, the Chow test is performed. This test allows us to detect a structural change at a given time. In this case, the mobility restrictions started on March 23rd, 2020, or day of the year (DOY) 83. Therefore, the test is performed for this day. The results of the Chow test for DOY 83 are shown in **Table 2**. The $PM_{2.5}$ diurnal cycle is also influenced by working hours since many $PM_{2.5}$ anthropogenic sources are related to transportation and industrial activities. Finally, the ozone diurnal cycle is also influenced by the diurnal cycle of solar radiation.

At the 95% confidence level, a significant change in $PM_{2.5}$ concentrations was found in the semi-urban areas. Ozone concentrations also changed significantly in the three areas on that day.

3.5 National standard exceedances

The final part of the assessment of the effect of lockdown on air quality is done by looking at the number of exceedances of the National Standard (NOM) for air quality (NOM-025-SSA1-2014 until 2022 and NOM-025-SSA1-2021 after 2022). The NOM-025-SSA1-2014 stipulates that daily $PM_{2.5}$ concentrations should not exceed 45 μg/m^3 and PM_{10} concentrations should not exceed 75 μg/m^3. In addition, the hourly average of O_3 should not be greater than 0.090 parts per million (ppm) in order to be considered "good" air quality.

The number of exceedances of the NOM for PM and ozone concentrations during the wet and dry seasons from 2012 to 2022 at urban, rural, and semi-urban stations in Mexico City is shown in **Figure 6**.

In the 2020 dry season (when the mobility restrictions were enacted), the recommended PM_{10} concentration levels were exceeded five times in urban areas and once in rural areas. Fewer NOM exceedances were reported that year than in previous years. The recommended levels of $PM_{2.5}$ concentration were exceeded only once in semi-urban areas during the dry season. In terms of $PM_{2.5}$ concentrations during the dry season, air quality improved in urban and rural areas due to major policy

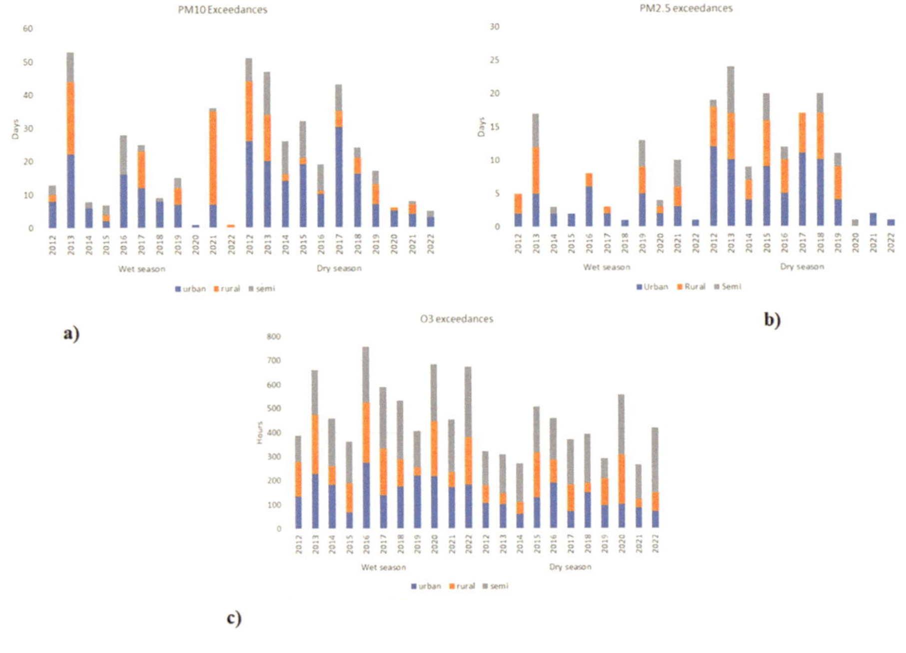

Figure 6.
(a) Number of exceedances of NOM for PM_{10}, (b) number of exceedances of NOM's limits for $PM_{2.5}$, and (c) number of exceedances of NOM for O_3. In all cases, the wet and dry seasons are shown for urban, rural, and semi-urban areas in GMC.

intervention. In contrast, the recommended levels of ozone were exceeded more often than in previous years, especially in semi-urban areas.

4. Discussion

The results of the analysis show that wind is the main environmental variable affecting PM and O_3 concentrations; therefore, long-range transport of pollutants can be a limitation of policies reducing anthropogenic sources of pollutants. A refinery complex is located around 80 km to the northeast of the city center in Tula, Hidalgo. It contributes to air pollution in GMC *via* long-range transport of pollutants, mainly PM_{10} [13, 24]. Also, high winds cause erosion of dry soils, especially in rural areas during the dry season [12]. Therefore, there is a positive significant relation between PM concentrations and wind speed in rural areas.

The diurnal cycles of PM_{10} concentration show that in the early morning, concentrations in semi-urban areas are higher than in rural and urban areas. As the working hours begin, PM_{10} concentrations increase in urban and rural areas. This has to do with transportation, industrial production, long-range transport of pollutants, soil erosion during the morning hours, and the boundary layer height cycle [25].

The ozone diurnal cycle is affected by solar radiation and economic activity since ozone precursors are related to vehicle emissions and solar radiation contributes to the formation of surface ozone [26].

The monthly averages show that the rainy season washes out pollutants, thus causing an improvement in air quality. Also, wet soils are less likely to be affected by wind erosion. Thus, some natural sources of PM and O_3 are reduced.

The trend analysis shows that in the long run, there was an effect of mobility restrictions (and halting or reduction of nonessential businesses and industries) in PM concentrations. However, ozone increased during this period as previous studies in the same region have shown [9]. The short-term analysis done with the Chow test showed that at the start of the restrictions, there was a decrease in $PM_{2.5}$ concentrations in semi-urban areas and ozone in the whole GMC. The effect on the other areas was not statistically significant. Finally, in 2020, the NOM of air quality was exceeded a few times in $PM_{2.5}$ and PM_{10} during the lockdown; however, the limits of ozone were exceeded more times than in the 10 previous years.

5. Conclusion

To analyze the effect of the reduction of anthropogenic sources on PM and O_3 concentrations, it is also necessary to analyze the effect of winds and other meteorological variables. Furthermore, long-range transport of pollutants and natural sources of pollutants such as wildfires or dry soils caused by drought need to be taken into account to assess the effect of policy interventions aimed at reducing anthropogenic sources of air pollutants in Greater Mexico City. According to the results of this study, during the 2020 dry season in Mexico City, there was a statistically significant reduction of $PM_{2.5}$ and PM_{10} concentrations, but there was an increase in ozone.

Acknowledgements

The authors thank Tania Isabel Rodríguez Mosqueda for the map in **Figure 1**.

Conflict of interest

The authors declare no conflict of interest.

Author details

Jorge Méndez-Astudillo[1*], Ernesto Caetano[2] and Karla Pereyra-Castro[3]

1 Institute of Economic Research, National Autonomous University of Mexico, Mexico City, Mexico

2 Institute of Geography, National Autonomous University of Mexico, Mexico City, Mexico

3 Faculty of Electronic Instrumentation, Universidad Veracruzana, Xalapa, Mexico

*Address all correspondence to: jmendez.astu@comunidad.unam.mx

IntechOpen

© 2023 The Author(s). Licensee IntechOpen. This chapter is distributed under the terms of the Creative Commons Attribution License (http://creativecommons.org/licenses/by/3.0), which permits unrestricted use, distribution, and reproduction in any medium, provided the original work is properly cited.

References

[1] INEGI. Censo de población y vivenda Mexico City 2021. 2021. Available from: https://www.inegi.org.mx/programas/ccpv/2020/

[2] Ramírez-Velázquez MR, Martínez-Reséndiz J. La dimensión regional de la movilidad y su impacto en la contingencia ambiental de la Ciudad de México. In: Coll-Hurtado A, Sánchez-Salazar MT, Mendoza-Vargas H, del Pont Lalli RM, editors. La movilidad en la Ciudad de México. Mexico City: Universidad Nacional Autónoma de México; 2018. p. 39-54

[3] Bravo H, Roy-Ocotla G, Sánchez P, Torres R. Contaminación atmosférica por ozono en la zona metropolitana de la ciudad de México: Evolución histórica y perspectivas. Omnia. 1991;7(23):39-48

[4] Cortina-Januchs MG, Barrón-Adame JM, Vega-Corona A, Andina D, editors. Pollution alarm system in Mexico. Bio-Inspired Systems: Computational and Ambient Intelligence. In: 10th International Work-Conference on Artificial Neural Networks, IWANN 2009; Salamanca, Spain. 2009

[5] Lei W, de Foy B, Zavala M, Volkamer R, Molina LT. Characterizing ozone production in the Mexico City Metropolitan Area: A case study using a chemical transport model. Atmospheric Chemistry and Physics. 2007;**7**(5):1347-1366

[6] Liu Y, Song M, Liu X, Zhang Y, Hui L, Kong L, et al. Characterization and sources of volatile organic compounds (VOCs) and their related changes during ozone pollution days in 2016 in Beijing, China. Environmental Pollution. 2020;**2020**:257

[7] Velasco E, Márquez C, Bueno E, Bernabé RM, Sánchez A, Fentanes O, et al. Vertical distribution of ozone and VOCs in the low boundary layer of Mexico City. Atmospheric Chemistry and Physics. 2008;**8**(12):3061-3079

[8] García-Reynoso A, Jazcilevich A, Ruiz-Suárez LG, Torres-Jardón R, Suárez Lastra M, Reséndiz Juárez NA. Ozone weekend effect analysis in México City. Atmosfera. 2009;**22**(3):281-297

[9] Peralta O, Ortínez-Alvarez A, Torres-Jardón R, Suárez-Lastra M, Castro T, Ruíz-Suárez LG. Ozone over Mexico City during the COVID-19 pandemic. Science Total Environment. 2021;**761**

[10] Vega E, Reyes E, Sánchez G, Ortiz E, Ruiz M, Chow J, et al. Basic statistics of PM2.5 and PM10 in the atmosphere of Mexico City. Science Total Environment. 2002;**287**(3):177-201

[11] Chow JC, Watson JG, Edgerton SA, Vega E. Chemical composition of PM2.5 and PM10 in Mexico City during winter 1997. Science Total Environment. 2002;**287**(3):177-201

[12] Díaz-Nigenda E, Tatarko J, Jazcilevich AD, García AR, Caetano E, Ruíz-Suárez LG. A modeling study of Aeolian erosion enhanced by surface wind confluences over Mexico City. Aeolian Research. 2010;**2**:143-157

[13] Mendez-Astudillo J, Caetano E, Pereyra-Castro K. Synergy between the urban Heat Island and the urban Pollution Island in Mexico City during the dry season. Aerosol and Air Quality Research. 2022;**22**(8)

[14] Secretaria de Movilidad. Plan estratégico de movilidad de la Ciudad de México 2019. Mexico City; 2019

[15] DGCA. Red de meteorología y radiación solar (REDMET). Mexico City: Secretaría del Medio Ambiente; 2021. Available from: http://www.aire.cdmx.gob.mx/objetivos-redes/objetivos-monitoreo-calidad-aire-redmet.html

[16] DGCA. Objetivos del monitoreo de la calidad del aire de la red automática de monitoreo atmosférico RAMA. Mexico City: Secretaría del Medio Ambiente; 2021. Available from: http://www.aire.cdmx.gob.mx/objetivos-redes/objetivos-monitoreo-calidad-aire-rama.html

[17] Balasmeh OA, Babbar R, Karmaker T. Trend analysis and ARIMA modeling for forecasting precipitation pattern in Wadi Shueib catchment area in Jordan. Arabian Journal of Geosciences. 2019;**12**(27)

[18] Chow GC. Tests of equality between sets of coefficients in two linear regressions. Econometrica. 1960;**28**:591-605

[19] Cerrato-Álvarez M, Miró-Rodríguez C, Pinilla-Gil E. Effect of COVID-19 lockdown on air quality in urban and suburban areas of Extremadura, Southwest Spain: A case study in usually low polluted areas. Revista Internacional de Contaminación Ambiental. 2021;**37**:237-247

[20] Thomas J, Jainet PJ, Sudheer KP. Ambient air quality of a less industrialized region of India (Kerala) during the COVID-19 lockdown. Anthropocene. 2020;**2020**:32

[21] Rani S, Kumar R, Acharya P, Maharana P, Singh R. Assessing the spatial distribution of aerosols and air quality over the Ganga River basin during COVID-19 lockdown p-1. Remote Sensing Applications: Society and Environment. 2021;**2021**:23

[22] Liao T, Wang S, Ai J, Gui K, Duan B, Zhao Q, et al. Heavy pollution episodes, transport pathways and potential sources of PM2.5 during the winter of 2013 in Chengdu (China). Science of the Total Environment. 2017;**584-585**:1056-1065

[23] Huang D, Li Q, Wang X, Li G, Sun L, He B, et al. Characteristics and Trends of Ambient Ozone and Nitrogen Oxides at Urban, Suburban, and Rural Sites from 2011 to 2017 in Shenzhen, China. Sustainability. 2018;**2018**:10

[24] García-Escalante J, García-Reynoso J, Jazcilevich-Diamant A, Ruiz-Suárez L. The influence of the Tula, Hidalgo complex on the air quality of the Mexico City metropolitan area. Atmosfera. 2015;**27**:215-225

[25] García-Franco JL, Stremme W, Bezanilla A, Ruíz-Angulo A, Grutter M. Variability of the mixed-layer height over Mexico City. Boundary-Layer Meteorology. 2018;**167**:493-507

[26] Li K, Jacob DJ, Liao H, Shen L, Zhang Q, Bates KH. Anthropogenic drivers of 2013-2017 trends in summer surface ozone in China. PNAS. 2018;**116**(2):422-427

Chapter 10

The Use of Stable Isotopes to Identify Carbon and Nitrogen Sources in Mexico City $PM_{2.5}$ during the Dry Season

Diego López-Veneroni and Elizabeth Vega

Abstract

Stable nitrogen and carbon isotopes were used to trace the interaction between atmospheric particles < 2.5 mm in diameter ($PM_{2.5}$) with atmospheric physical variables and atmospheric chemical species in an urban environment. $PM_{2.5}$ were collected daily at two sites in Mexico City during three dry seasons during two-week periods. $PM_{2.5}$ varied between 10 and 70 μg/m^3, with the highest concentrations occurring during low-speed southerly winds and the lowest during high-speed easterly winds. Stable carbon isotope composition ($\delta^{13}C$) showed that the main carbon source of $PM_{2.5}$ included emissions from fossil fuel combustion, along with low-molecular-weight carbon emissions and suspended dust. Stable nitrogen isotope values ($\delta^{15}N$) in PM varied between −9.9 and 21.6‰. The most ^{15}N-enriched particles generally occurred during low wind speeds and correlated significantly with hourly averaged ambient NO_x and NO_2 concentrations. Simultaneous samples from MER (commercial site) and XAL (industrial site) showed that $PM_{2.5}$ mass concentration was generally lighter and the carbon and nitrogen isotopic compositions were heavier at the commercial site relative to the industrial site. The $\delta^{15}N$ of $PM_{2.5}$ increased with the %N in $PM_{2.5}$ concordant with an isotopic fractionation during gas-to-particle condensation. Results suggest that wind speed, along with the nitrogen emission source, determines the nitrogen isotopic composition of $PM_{2.5}$.

Keywords: $PM_{2.5}$, $\delta^{13}C$, $\delta^{15}N$, Mexico City, ambient air

1. Introduction

Two major atmospheric contaminants are airborne particulate matter (PM) and nitrogen oxides (NO_x). PM from both natural and anthropogenic sources affect human health and air quality and, although they are size-regulated by national and international environmental standards, its control remains a challenge. PM has been associated with cardiovascular and respiratory diseases and life-expectancy reduction [1]. Because of its small size, $PM_{2.5}$ (PM < 2.5 mm in diameter) enter deep into the lungs and produce short- and long-time effects on the respiratory system, generating oxidative stress, systemic inflammation, and neuroinflammation, among other ailments [1, 2]. In contrast to PM_{10} (PM < 10 mm in diameter), which is usually composed of resuspended dust from unpaved roads, industrial emissions, agricultural activity, pollen, and bacteria, $PM_{2.5}$ is formed by combustion processes and the accretion of very small particles and/or condensation of gases on the surfaces of small particles [3]. $PM_{2.5}$ and PM_{10} contain numerous chemical compounds, such as carbon, sulfates, nitrates, and ammonium, among others, depending on the emission location. Primary particles are emitted directly from sources and can be formed through gas-to-particle conversion known as secondary particles. The secondary inorganic PM fraction is mainly composed of ammonium nitrate and ammonium sulfate whose precursors are emitted as NO_x, SO_2, and NH_3 and then converted to solids through chemical reactions (e.g., [4]).

In turn, nitrogen oxides (NO_x), especially nitrogen dioxide (NO_2), are strong atmospheric oxidants that enhance low atmospheric visibility, play a role in climate change, and are precursors of secondary contaminants [3]. As a result of anthropogenic activities, atmospheric nitrogen species and fluxes in urban environments far outweigh biogenic sources. For example, over 80% of nitrogen oxides (NO_x) are originated by the combustion of fossil fuels for transportation, electricity generation, and industrial activities. This contrasts with the 4% emissions by agricultural and biogenic sources [5]. Other nitrogen species also occur as gases or in particles in the atmosphere. NH_4^+ salts contribute to the long-range transport of acidic pollutants due to day-scale atmospheric lifetime and after deposition, they can contribute to forest decline and soil acidification. Ammonia and nitric acid are the main precursors of nitrate aerosols. Nitric acid is produced in the atmosphere as an additional reaction product of NO_2, from fossil fuel combustion, biomass burning, or from soil. NO_2 can oxidize and also react to form HNO_3 by pathways relied on the formation of NO_3^-.

Mexico City Metropolitan Area is characterized by being one of the largest megacities in the world. In addition to its more than 20 million inhabitants and a fleet of over 6 million vehicles, the air quality of the metropolitan area (which includes Mexico City and adjacent municipalities of the States of Hidalgo and Mexico) is affected by industrial and vehicular activity in the north and northeast sectors of the city and by biomass burning from a nearby agricultural activity, which is then transported into the urban area (e.g., [6]). The city lies on a high-altitude plateau (2240 m above sea level) and, except for the NE and SE sectors, is surrounded by mountain chains, which preclude an efficient pollutant dispersion (**Figure 1**). As such, in 2010, the $PM_{2.5}$ annual average concentration of 25 $\mu g/m^3$ was over two times higher than the recommended annual average of 10 $\mu g/m^3$ established by the World Health Organization (WHO), which clearly underscores the difficulty to reach the PM regulations [7]. Likewise, in 2010, the NO_2 annual average of 55 $\mu g/m^3$ was clearly greater than WHO's 40 $\mu g/m^3$. As a result of control enforcement policies during 2004–2008, $PM_{2.5}$ average concentration decreased by 27% [7], and in 2014, there was a decrease in PM of 11% with respect to 2013. Control measures, such as banning lead from gasoline, the use of natural gas, and mandatory vehicle emission inspection, have consistently decreased PM emissions.

Some studies on Mexico City's atmosphere have reported that $PM_{2.5}$ is mainly composed of 50% carbonaceous aerosol, suggesting an origin from incomplete combustion of fossil fuels and biomass burning, followed by sulfates, nitrates, ammonium, and geological material among others [8, 9]. The main secondary inorganic aerosol components in $PM_{2.5}$ mainly occurred as ammonium sulfate [$(NH_4)_2SO_4$] and ammonium nitrate [NH_4NO_3] due to the neutralization of atmospheric acids with gaseous

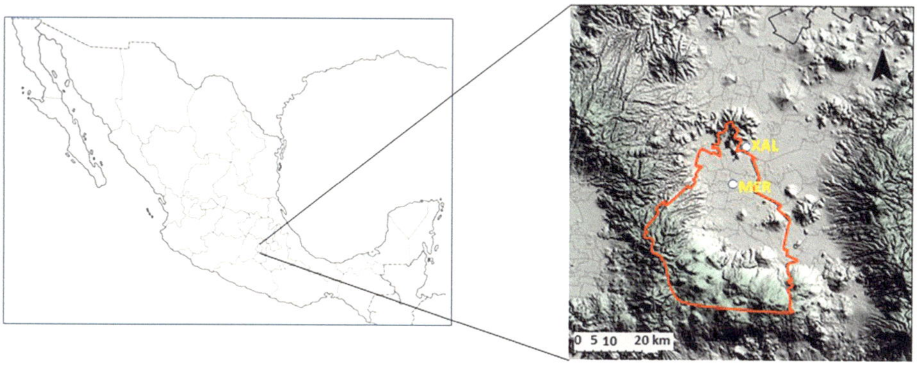

Figure 1.
Study zone where depicting XAL and MER sampling sites in Mexico City Metropolitan Area.

ammonia from agriculture, landfills, industries, biomass burning, and motor vehicles. In Mexico City, measurements made inside roadway tunnels showed a contribution of 8% of NH_4^+ emissions from on-road vehicles [10].

Stable nitrogen (^{15}N and ^{14}N) isotopes are valuable tools to trace the sources and transformations of airborne PM. Examples include tracing the sources of primary and secondary nitrogen in PM_{10} [11], nitrate accumulation in $PM_{2.5}$ [12], and NO_x contribution in nitrogen dry deposition [13]. On the other hand, stable carbon isotopes (^{13}C and ^{12}C) have been used in urban atmospheres to trace diesel and gasoline combustion, dust, soil and industrial emissions since these have distinct isotopic signatures [14–16]. Here, we use stable carbon isotopes and stable nitrogen isotopes of $PM_{2.5}$ from a data set to elucidate sources and chemical processes between the nitrogen in particles and several species of atmospheric nitrogen at two contrasting sites during three dry season periods in Mexico City. The sources of $PM_{2.5}$ are further studied using stable carbon isotopes. This chapter provides a baseline for $\delta^{15}N$ in Mexico City atmospheric particles as no previous data on N isotopes have been previously published.

2. Methodology

2.1 Sampling

During three two-week sampling campaigns (November 2003 and 2004, and March 2004), $PM_{2.5}$ samples were collected at two sites with different land use. **Figure 1** shows the two sampling sites, along with the orographic characteristics of the Mexico City Metropolitan Area. La Merced (MER) is an important commercial site with a major food market, located approximately 0.2 km west of downtown; the site is near to moderately traveled paved streets. Xalostoc (XAL) is located in the northeast of the city, adjacent to heavily traveled paved and unpaved roads where old and new gasoline and diesel vehicles transit in an industrialized area; this site is 5 km west of the dry Lake Texcoco, thus resuspended dust from the dry lake bed may affect this site.

Particle samples were collected with battery-powered MiniVol portable $PM_{2.5}$ samplers (Airmetrics) operating with a volume intake of air of 5 L/min for 24 h. Before sampling, the filtration volume of the samplers was previously calibrated at standard conditions of temperature and pressure. Teflon membranes were used to measure mass by gravimetry. Details of the equipment and sampling procedures can be found elsewhere [9].

2.2 $PM_{2.5}$ concentration and composition measurements

$PM_{2.5}$ mass was determined by duplicate samples collected on 47 mm Teflon filters, which were weighted in a microbalance prior to and after sampling.

Ions were collected on quartz filters and extracted in deionized water by sonication. Ions were analyzed by high-performance liquid chromatography. Nitrate, sulfate, and ammonium were analyzed by ion exchange chromatography. Organic carbon and elemental carbon were analyzed by thermal–optical reflectance.

2.3 Stable nitrogen and carbon isotopes measurements

Stable carbon and nitrogen isotopes of $PM_{2.5}$ were collected in pre-combusted GF/F filters, which were placed in the MiniVol samplers. After sampling, filters were freeze-dried to remove any water or humidity in the sample, placed in aluminum vials, and inserted into a combustion column. Generated gases (CO_2 and N_2) were analyzed in a MAT-Finnigan 252 stable isotope mass spectrometer to quantify the $^{13}C/^{12}C$ and $^{15}N/^{14}N$ isotopes in each sample. Standards used for instrument calibration were acetanilide, methionine, and urea, all with a known isotopic value. An acetanilide substandard was run for every 10 samples. Carbon and nitrogen isotopes are referred to as VPDB and Vienna air nitrogen standards, respectively.

2.4 Meteorological and ambient air data

Ancillary ambient data were obtained from the RAMA and REDMET data sets. RAMA (Spanish acronym for automatic ambient monitoring network; [17]) measures criteria atmospheric pollutants (PM_{10}, CO, CO_2, NOx, NO_2, SO_2, and O_3) at 15 locations within Mexico City on an hourly basis. At MER, $PM_{2.5}$ hourly data are also measured by RAMA. Wind velocity and direction, relative humidity, and temperature measured on an hourly basis were obtained from the REDMET (meteorological network) database [17]. For each sampling period, the hourly data were averaged, except for wind direction where the hourly mode was used.

3. Results and discussion

3.1 Meteorological characterization

Table 1 gives the basic statistics of meteorological variables during the three periods of study. On average, November 2003 was colder than the other sampling periods. In turn, March 2004 was drier and had the highest wind speeds. November 2004 was characterized by low wind speeds. During the first two sampling periods, the most frequent wind direction was from the NW-NE sectors. In November 2004, both XAL and MER sites were affected by E-SE winds

The diurnal variation of the meteorological parameters during the study is shown in **Figure 2**. The hourly data at each sampling period and site is an average of 14 days. In all three sampling periods and the two sites, early morning was characterized by low temperatures at 6 h, which increased consistently to a maximum at 16 h to then

		Temperature (°C)			Humidity (%)			Wind speed (m/s)			Wind direction (°)
Date	Site	Average	Min	Max	Average	Min	Max	Average	Min	Max	Mode
Nov 2003	MER	13.96	3.40	24.50	54.27	24.00	90.00	1.18	0.04	3.86	30–60, 0–30
	XAL	13.90	4.00	24.00	55.62	21.00	84.00	1.65	0.07	6.52	30–60, 0–30
Mar 2004	MER	17.75	8.80	29.1	46.83	19.00	80.00	1.61	0.15	5.07	0–30, 330–360
	XAL	16.87	8.60	27.10	53.74	25.00	84.00	2.18	0.02	7.49	0–30, 30–60
Nov 2004	MER	17.02	8.60	26.10	54.26	18.00	94.00	1.14	0.17	3.69	60–90, 90–120, 30–60
	XAL	16.40	8.80	25.20	53.52	20.00	84.00	0.92	0.04	4.80	120–150, 330–360

Table 1.
Basic statistics of meteorological variables during the three periods of study.

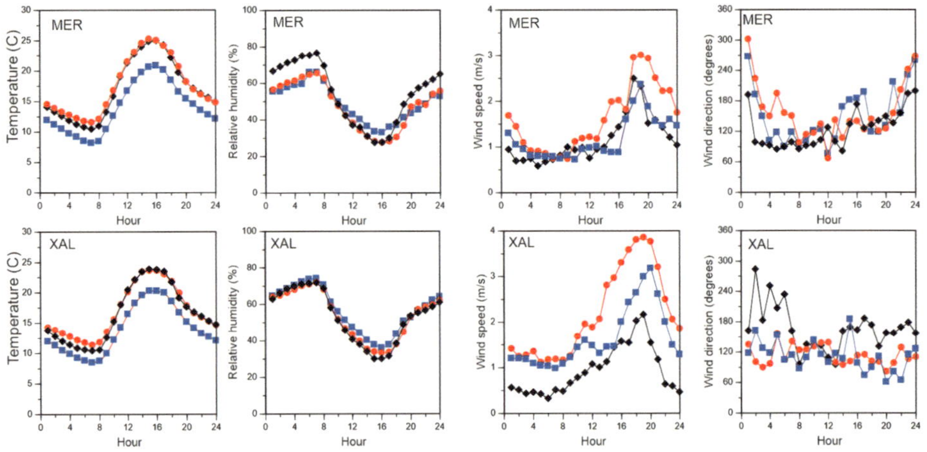

Figure 2.
Hourly average of temperature, relative humidity, wind speed, and wind direction at MER and XAL during November 2003 (blue squares), March 2004 (red dots), and November 2004 (black diamonds). Each symbol is the hourly average of 14 days at each sampling site from the Mexico City Meteorological Network (REDMET).

decrease again. In turn, the most humid conditions were present at 6 h and decreased to a minimum of 30% at 16 h. Wind speeds were usually low during the morning hours and increased to a maximum around 18 h, which favors contaminant dispersion from the valley. At MER, low wind speeds were consistently slow until 14 h, in contrast to speeds at XAL where wind speed increased from 6 h to maximum speeds at 18 h. The lowest daily average wind speeds were significantly lower at XAL during November 2004. In general, westerly winds were present at MER between 22 h and the next day 2 h. On average, easterly winds were dominant throughout the rest of the day at both locations.

3.2 $PM_{2.5}$ concentrations

Scatterplots of the daily $PM_{2.5}$ concentrations collected with the MiniVol samplers were compared with the respective average daily PM_{10} concentrations from hourly measurements from the RAMA stations at XAL and MER (**Figure 3**). At MER, hourly measurements of the MiniVol $PM_{2.5}$ data were also compared with RAMA $PM_{2.5}$ data. The scatterplots show a high correlation ($R^2 > 0.9$) between $PM_{2.5}$ from our MiniVols and the RAMA data and show that our daily samples are representative of the average daily $PM_{2.5}$ concentrations [18]. The high correlation between $PM_{2.5}$ and PM_{10} further suggests a common source or formation mechanism of these particles. López-Veneroni [14] showed similar correlations in concentrations and carbon isotope compositions for simultaneous $PM_{2.5}$ and PM_{10} samples in Mexico City.

The average diurnal variations of PM_{10} at XAL and MER, and of $PM_{2.5}$ at MER from the RAMA stations during the period of study are shown in **Figure 4**. These data provide insight into the time of the major particle accumulation throughout the day. Morning $PM_{2.5}$ concentrations were around 30 mg/m^3 and peaked to 70 μg/m^3 at 10 h. Concentrations then decreased until 18 h and remained constant at 30 μg/m^3 throughout the night and early morning. In contrast, the diurnal variation of PM_{10} showed two major peaks at 8 h and 18 h, which probably result in particle accumulation during traffic rush hours. During the afternoon, the increase in wind speeds dispersed these pollutants.

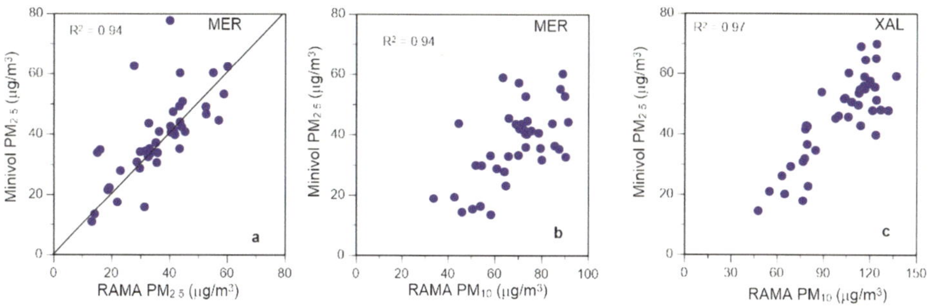

Figure 3.
Scatterplots of (a): $PM_{2.5}$ samples collected daily by MiniVol samplers vs. 24 h averaged $PM_{2.5}$ samples from Mexico City atmospheric monitoring automatic network (RAMA) at MER. (b): $PM_{2.5}$ samples collected daily by MiniVol samplers vs. 24 h averaged PM_{10} from RAMA at MER. (c): $PM_{2.5}$ samples collected daily by MiniVol samplers vs. 24 h averaged PM_{10} from RAMA at XAL.

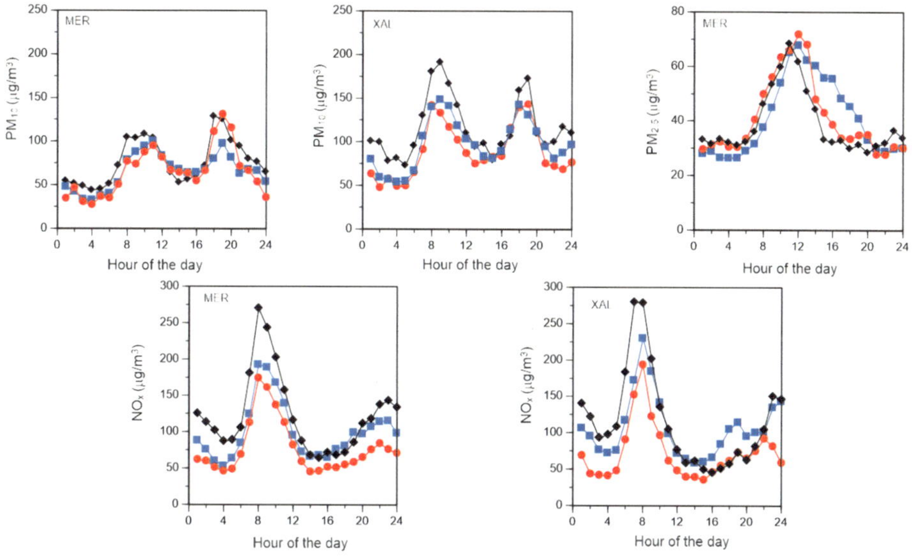

Figure 4.
Average diurnal variations of PM_{10} concentrations and NO_x concentrations at MER and XAL from Mexico City atmospheric monitoring automatic network (RAMA), and $PM_{2.5}$ diurnal variations at MER from RAMA, during November 2003 (blue squares), March 2004 (red dots), and November 2004 (black diamonds).

Average $PM_{2.5}$ concentrations for the two sites and three sampling periods are given in **Table 2**. In November 2003 and May 2004, concentrations were similar in both sites. In November 2004, $PM_{2.5}$ concentrations were significantly lower at MER relative to XAL, in accordance with the industrial activity of this site [9, 10] and low wind speeds, which preclude pollutant dispersion. Average concentrations were below 65 µg/m^3 (the Mexican Health Standard Norm at the time of the study) at all sampling stations and dates (the current allowable maximum is 41 µg/m^3, [7]). In the two sites, $PM_{2.5}$ concentrations were highest during weekdays (Monday to Friday) relative to the weekends (t-test, $p < 0.01$).

Average concentrations of total carbon, nitrate, and ammonium for each sampling period were also similar at MER and XAL. **Figure 5** shows that wind speed is

Date	Site	PM2.5 (μg/m³)		δ¹³C (‰)		δ¹⁵N (‰)		NH4 (μg/m³)		NO3 (μg/m³)		Total carbon (μg/m³)	
		Ave (stdev)	**Range**	**Ave (stdev)**	**Range**	**Ave (stdev)**	**Range**	**Ave (stdev)**	**Range**	**Ave (stdev)**	**Range**	**Ave (stdev)**	**Range**
Nov 2003	MER	39.40 (12.93)	18.69 to 60.07	−25.69 (1.93)	−27.43 to −21.98	9.84 (5.32)	−2.27 to 18.26	3.49 (2.07)	1.23 to 8.94	5.60 (4.14)	1.75 to 17.18	**13.47** (4.23)	7.83 to 21.00
	XAL	46.89 (15.16)	20.93 to 69.90	−25.28 (1.46)	−27.00 to −23.09	10.02 (6.83)	−2.72 to 21.64	3.68 (2.28)	0.85 to 9.39	5.62 (3.88)	1.13 to 15.49	**20.68** (7.17)	10.59 to 35.13
March 2004	MER	28.96 (11.44)	13.33 to 43.61	−27.02 (0.53)	−27.95 to −26.28	**5.45** (4.56)	−2.19 to 15.43	2.89 (1.67)	0.96 to 6.17	3.37 (2.43)	1.09 to 9.54	12.70 (3.45)	7.01 to 18.41
	XAL	37.14 (14.89)	14.29 to 60.17	−28.95 (5.56)	−41.26 to −24.02	**0.93** (5.41)	−7.51 to 10.50	2.77 (1.77)	0.53 to 5.99	3.66 (2.66)	0.42 to 8.66	16.26 (5.83)	5.70 to 23.76
Nov 2004	MER	**39.60** (9.96)	23.01 to 58.84	**−20.62** (3,11)	−25.65 to −15.31	5.97 (8.47)	−9.90 to 19.43	3.42 (2.12)	1.55 to 9.65	5.11 (2.24)	1.99 to 9.22	20.11 (3.89)	13.89 to 28.11
	XAL	**52.89** (6.94)	42.43 to 64.99	**−22.76** (0.90)	−23.89 to −20.40	2.67 (2.23)	−0.61 to 8.51	3.23 (1.61)	1.59 to 7,97	5.37 (1.81)	3.56 to 10.57	21.99 (3.59)	14.09 to 28.46

Significant differences between sites for each sampling season are given in bold (t-test < 0.05).

Table 2.
Average and standard deviation of carbon and nitrogen parameters of $PM_{2.5}$ in XAL and MER during November 2003, March 2004, and November 2004.

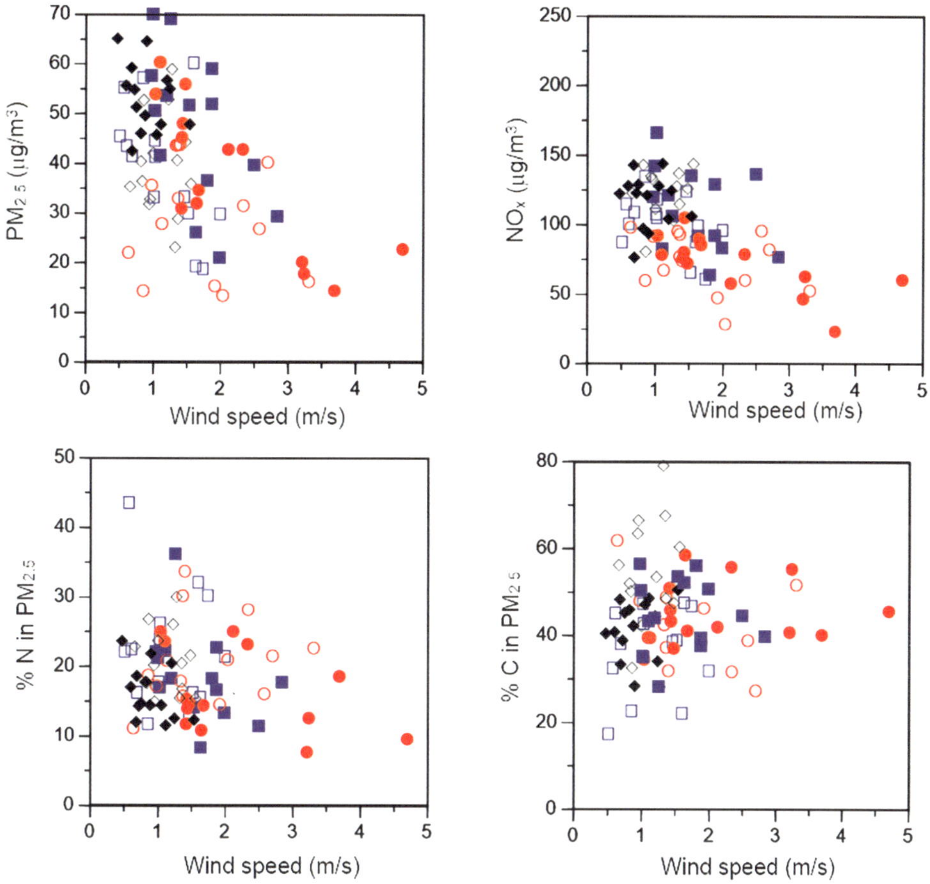

Figure 5.
Scatterplots of wind speed vs. $PM_{2.5}$ concentrations, averaged NO_x concentrations from Mexico City atmospheric monitoring automatic network (RAMA), % N in $PM_{2.5}$, and % C in $PM_{2.5}$ for samples collected at MER (closed symbols) and XAL (open symbols) during November 2003 (blue symbols), March 2004 (red symbols), and November 2004 (black symbols).

important in determining the $PM_{2.5}$ concentration and carbon and nitrogen composition. As wind speeds decrease, particle concentrations increase and suggest that the emitted (primary) or coalesced (secondary) particles increase when no mechanism disperses them. Likewise, the percentage of N and NO_3– concentration in $PM_{2.5}$ also increased at low wind speeds, in accordance to the gas-to-particle conversion of nitrogen compounds [4]. By contrast, the percentage of the total carbon is increased along with an increase in wind speeds and a decrease in particle concentrations, and this suggests that at high wind speeds low particle concentrations appear to be primarily composed of direct carbon emissions.

3.3 Stable carbon and nitrogen isotopes of $PM_{2.5}$

Figure 6 gives the frequency histograms of stable carbon and nitrogen isotopes in $PM_{2.5}$ collected at MER and XAL during the three sampling campaigns. Averages and ranges are given in **Table 2**.

The majority of $\delta^{13}C$ values fell in the –27 to –25‰ range, with an important number of data points skewed to more positive values (–24 to –15‰). A few samples

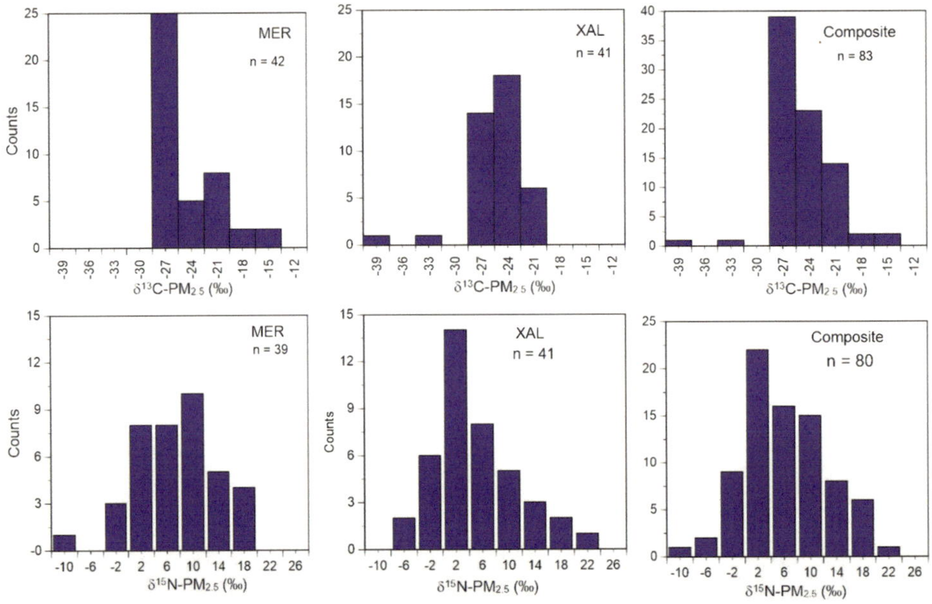

Figure 6.
Upper panel: histograms of $\delta^{13}C$-$PM_{2.5}$ (‰) at MER, XAL, and all data. Lower panel: histograms of $\delta^{15}N$-$PM_{2.5}$ (‰) at MER, XAL, and all data.

at XAL were lighter than −28‰. In contrast, during November 2004 at MER, most $\delta^{13}C$ values were heavier than −22‰. This range of isotopic compositions for bulk carbon contrasts with data from a previous study in Mexico City, where most $\delta^{13}C$ in $PM_{2.5}$ ranged between −26 and −24‰ [14]. According to the carbon isotope composition of the different potential sources in Mexico City, the predominant carbon source in $PM_{2.5}$ is the emissions of fossil fuels [14]. The extreme isotopic values at XAL show

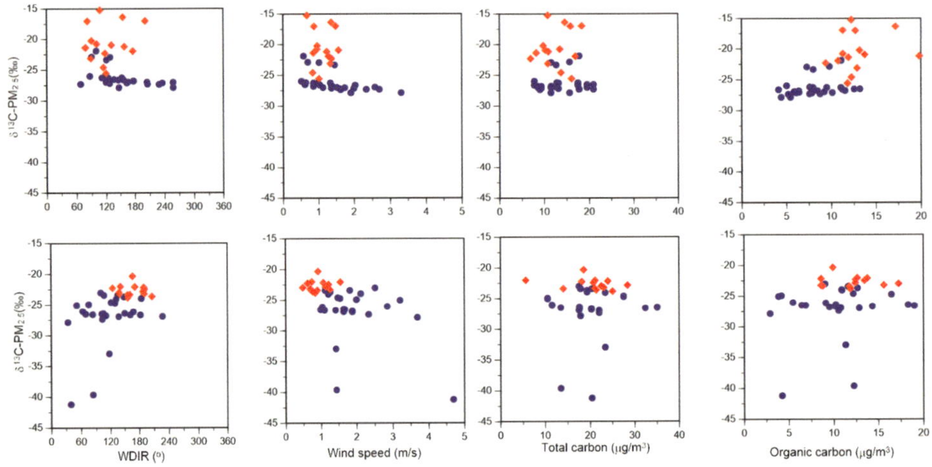

Figure 7.
Upper panel: scatterplots of $\delta^{13}C$-$PM_{2.5}$ vs. wind direction, wind speed, total carbon, and organic carbon at MER. Lower panel: scatterplots of $\delta^{13}C$-$PM_{2.5}$ vs. wind direction, wind speed, total carbon, and organic carbon at XAL. Blue symbols denote samples collected during November 2003 and March 2004. Red symbols denote samples collected during November 2004.

that during March 2004, light hydrocarbons (such as methane and propane) were the predominant emissions. In contrast, the relatively enriched δ^{13} C values at MER during November 2004 reflect the emission of particles of geological origin.

Scatterplots of $\delta^{13}C$ values vs. wind direction, wind speed, organic carbon, and total carbon are depicted in **Figure 7**. The enriched ^{13}C values during November 2004 are associated with low-speed E-SE winds and suggest organic-rich, soil-derived carbon. In contrast, the carbon content of $PM_{2.5}$ in the other two sampling periods at MER is related to emissions from fuel combustion. At XAL, the heaviest $\delta^{13}C$ values are associated with low-speed winds and organic carbon-enriched particles. The ^{13}C-depleted samples apparently originate from SE winds.

$\delta^{15}N$-$PM_{2.5}$ values spanned between −9.9 and +21.6‰ (**Table 2**). Although the average $\delta^{15}N$ composition between sites for a given sampling period was similar (except for March 2004), the frequency histogram of $\delta^{15}N$ in $PM_{2.5}$ shows a different distribution. At XAL, most $\delta^{15}N$ values fell in the 2‰ bin, with over 50% of the

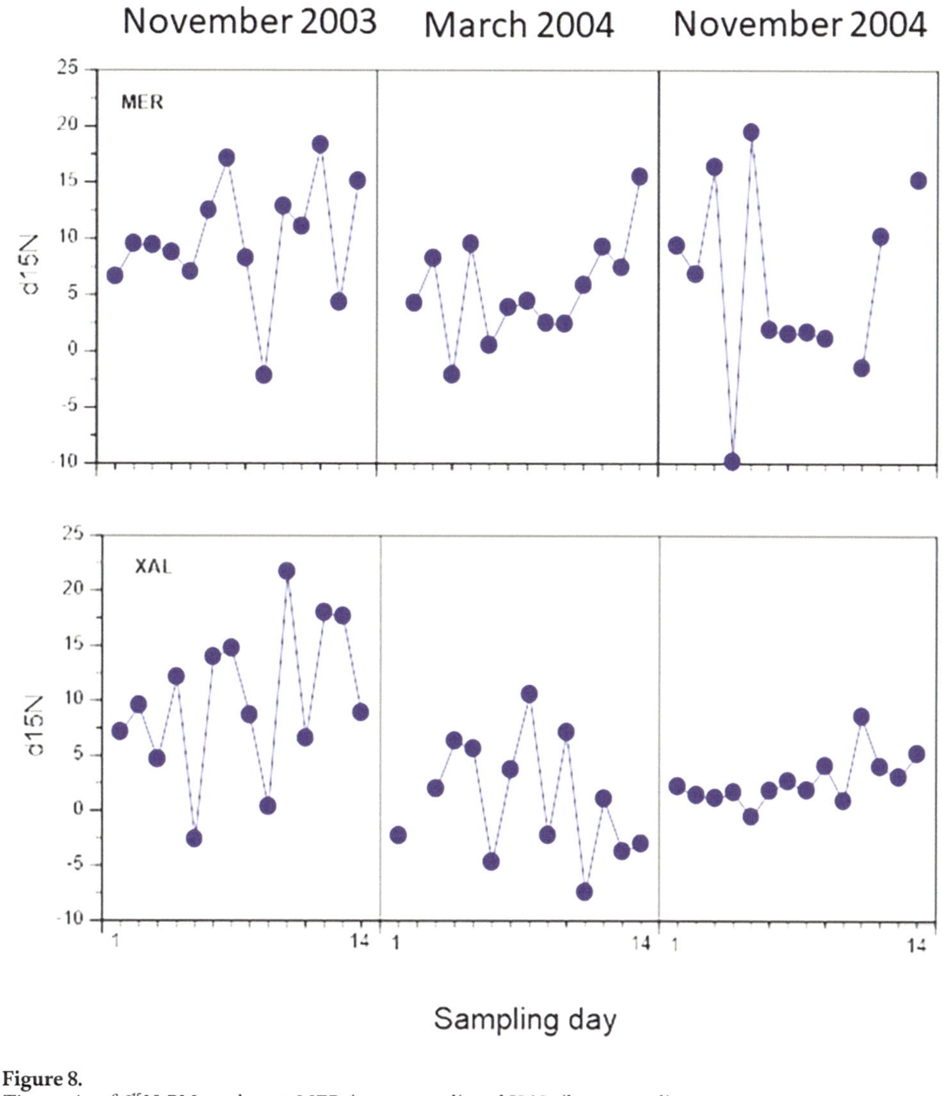

Figure 8.
Time series of $\delta^{15}N$-$PM_{2.5}$ values at MER (upper panel) and XAL (lower panel).

data points between −2 and 6‰. In contrast, the $\delta^{15}N$-$PM_{2.5}$ distribution at MER had a wider distribution, with most values between 2 and 10‰. The time series of $\delta^{15}N$-$PM_{2.5}$ at the two sites shows that at XAL, values were always lower than at

Location	Particle size	Season	Average + std. deviation	Range	Comments	Reference
Paris, France	PM10	winter	10.0 ± 3.4	5.3 to 15.9		[15]
Paris, France	PM10	summer	10.8 ± 3.4	5.9 to16.1		[15]
Cienfuegos, Cuba	PM10	urban	9.2 ± 4.4	1.5 to 19.1		[19]
Seoul Korea	PM2.5 NH4	warm	16.4 ± 2.8			[20]
Seoul Korea	PM2.5 NH4	cold	4.0 ± 6.1			[20]
Seoul Korea	PM2.5 NO3	summer	−0.7 ± 3.3			[21]
Seoul Korea	PM2.5 NO3	winter	3.8 ± 3.7			[21]
Shijiazhuang, China	PM2.5 NO3	warm months		−11.8 to 13.8	No differences in $\delta^{15}N$-NO_3 between polluted and non-polluted days	[12]
Shijiazhuang, China	PM2.5 NO3	cold months		−0.7 to 22.6		[12]
Northeast cities, USA	PM2.5 NO3					[13]
New Delhi, India	PM2.5 bulk	Spring summer	12.3 ± 4.6			[22]
New Delhi, India	PM2.5 bulk	Post-monsoon	7.7 ± 4.1			[22]
New Delhi, India	PM2.5 bulk	winter	7.3 ± 6.6			[22]
Beijing, China	PM2.5 bulk	winter	11.97 ± 1.79	8.68 to 14.50	Sampling in haze and non-haze conditions	[23]
Chennai, India	PM2.5 bulk	summer	22.2 ± 4.3	19.3 to 25.2		[24]
Chennai, India	PM2.5 bulk	winter	25.5 ± 2.4	18.0 to 27.8		[24]
Mexico City, Mexico	PM2.5 bulk	dry cold	4.6 ± 6.4	−7.5 to 21.6	XAL-industrial site	This study
Mexico City, Mexico	PM2.5 bulk	dry cold	7.2 ± 6.4	−9.9 to 19.4	MER-residential/ commercial site	This study

Table 3.
$\delta^{15}N$ values in $PM_{2.5}$ and PM_{10} in megacities.

MER (**Figure 8**). Furthermore, $\delta^{15}N$ for $PM_{2.5}$ at XAL during the low wind speeds of November 2004, was nearly constant during the sampling period.

A comparison between $\delta^{15}N$ for $PM_{2.5}$ in Mexico City with those of other megacities is shown in **Table 3**. The average values for XAL and MER are similar to the averages of New Delhi [22] and several months in Shijiazhuang [12], but lighter than for Paris [11], Beijing [23], and the warm months of New Delhi [22] and Seoul [21]. These differences can be attributed to different sources of NOx emissions [13, 23].

Scatterplots of XAL and MER $\delta^{15}N$ values of $PM_{2.5}$ samples vs. wind speed and averaged air quality data are given in **Figure 9**. The figure shows that as windspeed increases, the nitrogen isotopic composition becomes ^{15}N-depleted, while at low wind speeds, $PM_{2.5}$ particles are isotopically enriched. In turn, $\delta^{15}N$ values are positively correlated with average daily NO_x ambient concentrations. At low wind speeds, particles are dispersed and the isotopic composition should reflect primary emissions. This is consistent with low $\delta^{15}N$ values of fossil fuels. In turn, the NOx- $\delta^{15}N$-$PM_{2.5}$ positive correlation shows that the gas-to-particle condensation fractionates the condensed nitrogen, leaving isotopically enriched N in the particles.

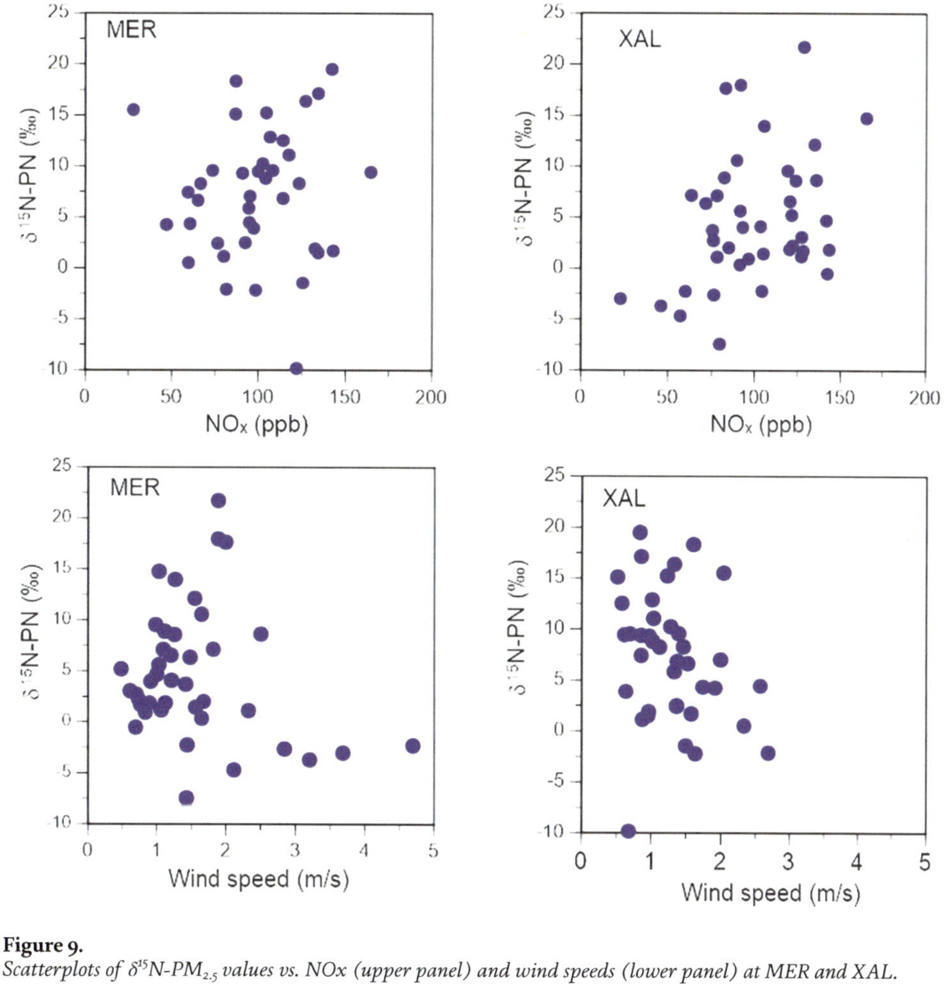

Figure 9.
Scatterplots of $\delta^{15}N$-$PM_{2.5}$ values vs. NOx (upper panel) and wind speeds (lower panel) at MER and XAL.

4. Conclusions

Potential N and C sources of $PM_{2.5}$ in an industrial sector and a commercial/residential zone of Mexico City were evaluated during three dry season periods (March and November) using stable nitrogen and carbon isotopes. Daily $PM_{2.5}$ concentrations ranged between 10 and 70 μg/m^3 and were inversely correlated to wind speed. The percentage of N in $PM_{2.5}$ also increased at low wind speeds and suggests gas-to-particle conversion of nitrogen compounds. In contrast, the % C increased with high wind speeds and low $PM_{2.5}$ concentrations reflecting direct carbon emissions. Based on $\delta^{13}C$ values, the principal carbon sources of $PM_{2.5}$ are fossil fuel emissions, although geological material is an important component when easterly winds resuspend dust from the nearby Texcoco dry lakebed. $\delta^{15}N$-$PM_{2.5}$ values ranged between −9.9 and 21.6 ‰, similar to values from other megacities. $\delta^{15}N$ values were consistently lower at the industrial site and suggest isotopic fractionation of NOx emissions during particle accretion. The $\delta^{15}N$-$PM_{2.5}$ values presented here provide a nitrogen isotope baseline for Mexico City airborne fine particles as no previous data have been collected for this megacity.

Acknowledgements

Research presented in this chapter was funded by the Instituto Mexicano del Petróleo under project Y.00169, and CONACYT-SEMARNAT project 2002-CO1-0296: *Análisis de la fracción orgánica, metales, distribución de tamaño y propiedades ópticas de partículas suspendidas en la atmósfera de la ZMCM*. We acknowledge Dr. Luis A. Cifuentes and Brian Jones (Stable Isotope Ecology Laboratory, Department of Oceanography, Texas A&M University, U.S.A.) for the isotope analyses.

Conflict of interest

The authors declare no conflict of interest.

Author details

Diego López-Veneroni[1] and Elizabeth Vega[2*]

1 Independent Researcher, Mexico

2 Instituto de Ciencias de la Atmósfera y Cambio Climático, UNAM, Mexico

*Address all correspondence to: evega@atmosfera.unam.mx

IntechOpen

© 2022 The Author(s). Licensee IntechOpen. This chapter is distributed under the terms of the Creative Commons Attribution License (http://creativecommons.org/licenses/by/3.0), which permits unrestricted use, distribution, and reproduction in any medium, provided the original work is properly cited.

References

[1] Pope CA III, Ezzati M, Dockery DW. Fine-particulate air pollution and life expectancy in the United States. New England Journal of Medicine. 2009;**360**:376-386

[2] Calderón-Garcidueñas L, Kulesza RJ, Doty RL, D'Angiulli A, Torres-Jardón R. Megacities air pollution problems: Mexico City Metropolitan Area critical issues on the central nervous system pediatric impact. Environmental Research. 2015;**137**:157-169

[3] WHO (World Health Organization). Air Quality Guidelines. Global Update 2005. 2006. World Health Organization. www.euro.who.int

[4] Yao X, Lau APS, Fang M, Chan CK, Hu M. Size distribution and formation of ionic species in atmospheric particulate pollutants in Beijing, China: 1- inorganic ions. Atmospheric Environment. 2003;**37**:2991-3000

[5] Shaw S, Van Heyst B. Nitrogen oxide (NOx) emissions as an indicator for sustainability. Environmental and Sustainability Indicators. 2022;**15**:100188

[6] Carabali G, Villanueva-Macías J, Ladino LA, Álvarez-Ospina H, Raga GB, Andraca-Ayala G, et al. Characterization of aerosol particles during a high pollution episode over Mexico City. Scientific Reports. 2021;**11**:22533

[7] SEDEMA (Secretaría del Medio Ambiente de la Ciudad de México). Calidad del Aire de la Ciudad de México. Informe 2018. Dirección General de Calidad del Aire, Dirección de Monitoreo de Calidad del Aire. 2020

[8] Querol X, Pey J, Minguillón MC, Pérez N, Alastuey A, Viana M, et al. PM speciation and sources in Mexico during the MILAGRO-2006 Campaign. Atmospheric Chemistry and Physics. 2008;**8**:111-128

[9] Vega E, Escalona S, Cervantes A, López-Veneroni D, González Avalos E, Sánchez Reyna G. Chemical composition of fine particles in Mexico City during 2003-2004. Atmospheric Pollution Research. 2011;**2**:477-483

[10] Vega E, Reyes E, Ruiz H, García J, Sánchez G, Martínez-Villa G, et al. Analysis of $PM_{2.5}$ and PM_{10} in the atmosphere of Mexico City during 2000-2002. Journal of the Air & Waste Management Association. 2004;**54**:786-798

[11] Widory D. Nitrogen isotopes: Tracers of origin and processes affecting PM_{10} in the atmosphere of Paris. Atmospheric Environment. 2007;**41**:2382-2390

[12] Luo L, Zhu R-G, Song C-B, Peng J-F, Guo W, Liu Y, et al. Change in nitrate accumulation mechanisms as $PM_{2.5}$ levels increase on the North China Plain: A perspective from dual isotopic compositions of nitrate. Chemosphere. 2021;**263**:127915

[13] Elliott EM, Kendall C, Boyer EW, Burns DA, Lear GG, Golden HE, et al. Dual nitrate isotopes in dry deposition: Utility for partitioning NOx source contributions to landscape nitrogen deposition. Journal of Geophysical Research. 2009;**114**:G04020

[14] López-Veneroni D. The stable carbon isotope composition of $PM_{2.5}$ and PM_{10} in Mexico City Metropolitan Area air. Atmospheric Environment. 2009;**43**:4491-4502

[15] Gorka M, Jedrysek MO. Solid atmospheric particles and wet precipitations in Wroclaw (SW Poland): Mineralogical and isotopic preliminary studies. Mineralogical Society of Poland – Special Papers. 2004;**24**:179-182

[16] Tanner RL, Miguel AH. Carbonaceous aerosol sources in Rio de Janeiro. Aerosol Science and Technology. 1989;**10**:13-223

[17] SEDEMA (Secretaría del Medio Ambiente de la Ciudad de México). Inventario de Emisiones de la Zona Metropolitana del Valle de México 2018. Dir. Gral de Calidad del Aire, Dirección de Proyectos de Calidad del Aire. Ciudad de México. 2021

[18] Hernández-López AE, Martin del Campo JM, Múgica Álvarez V, Valle-Hernández BL, Mejía-Ponce LV, Pineda-Santamaría JC, et al. A study of $PM_{2.5}$ elemental composition in southwest Mexico City and development of receptor models with positive matrix factorization. Revista Internacional de Contaminación Ambiental. 2021, 2021;**37**:67

[19] Morera-Gómez Y, Santamaría JM, Elustondo D, Alonso-Hernández CM, Widory D. Carbon and nitrogen isotopes unravel sources of aerosol contamination at Caribbean rural and urban coastal sites. Science of the Total Environment. 2018;**642**:723-732

[20] Lim S, Hwang J, Lee M, Czimczik CI, Xu X, Savarino J. Robust evidence of ^{14}C, ^{13}C, and ^{15}N analyses indicating fossil fuel sources for total carbon and ammonium in fine aerosols in Seoul Megacity. Environmental Science and Technology. 2022;**56**:6894-6904

[21] Lim S, Lee M, Savarino J, Laj P. Oxidation pathways and emission sources of atmospheric particulate nitrate in Seoul: Based on $\delta^{15}N$ and $D^{17}O$ of $PM_{2.5}$. Atmospheric Chemistry and Physics. 2022;**22**:5099-5115

[22] Sawlani R, Agnihotri R, Sharma C. Chemical and isotopic characteristics of PM2.5 over New Delhi from September 2014 to May 2015; Evidences for synergy between air-pollution and meteorological changes. Science of the Total Environment. 2021;**763**:142966

[23] Guo X, Li C, Tang L, Briki M, Ding H, Ji H. Sources of organic matter (PAHs and n-alkanes) in $PM_{2.5}$ of Beijing in haze weather analyzed by combining the C-N isotopic and PCA-MLR analyses. Environmental Science: Processes and Impacts. 2016;**18**:314-322

[24] Pavuluri CM, Kawamura K, Tachibana E, Swaminathan T. Elevated nitrogen isotope ratios of tropical Indian aerosols from Chennai: Implication for the origins of aerosol nitrogen in South and Southeast Asia. Atmospheric Environment. 2010;**44**:3597-3604

Chapter 11

Analysis of Temporal Lag in the Impact of Air Quality on the Health of Children, in Barreiro

João Garcia, Rita Cerdeira and Luís Coelho

Abstract

The aim of this work was to study the impact of temporal lag between the exposition to air pollutants and the children admitted to the emergency room of Hospital Nª Srª Rosário pediatric service, in Barreiro, Portugal, with symptoms of respiratory problems. The two variables were recorded by the medical staff and by an air quality monitoring station, in the same periods. From the results, a moderate correlation between different symptoms of respiratory diseases (sdr, cough, and asthma) and pollutants was found, reaching maximum values after temporal lags of 2 to 6 days. The strongest correlation for lag 0 (consequences on the same day) rises for the symptomatology of asthma, reaching the highest values for COmax ($\rho = 0.26$) and COpeak ($\rho = 0.25$). Also, an important correlation was found for NO_x, NO_x med and peak NO_x ($\rho = 0, 21$). The correlation with PM_{10} shows an unrepresentative value ($\rho = 0.09$), being negative for O_3max ($\rho = -0.23$) and O_3peak ($\rho = -0.22$), as well as for SO_2med ($\rho = -0.12$). Considering temporal lags of 1 to 8 and 15 days, overall, the maximum correlations between symptoms and NO, NO_2, NO_x, CO, and PM_{10} occur after temporal lags of 2 to 6 days, being constant or negative to SO_2 and O_3.

Keywords: air quality, children's health, correlation, temporal lag, respiratory distress

1. Introduction

There are several air pollutants that can have a significant impact on human health [1, 2], including particulate matter (PM) which can be inhaled into the lungs and cause respiratory and cardiovascular problems, as well as increase the risk of lung cancer [3–5]. Nitrogen oxides (NO_x) can irritate the lungs and aggravate asthma and other respiratory problems. It can also contribute to the formation of ground-level ozone, which can cause respiratory and cardiovascular problems [6]. Sulfur dioxide (SO_2) can irritate the respiratory system and aggravate asthma and other respiratory problems [7]. Carbon monoxide (CO) can be deadly in high concentrations, but even at low concentrations, it can cause headaches, dizziness, and nausea. Also, Ozone (O_3) can cause respiratory and cardiovascular problems, especially in people with preexisting conditions such as asthma [8, 9]. These health effects of these pollutants can vary depending on factors such as the concentration and duration of exposure, age, and preexisting health conditions [10]. It is known that children, the elderly, and the people with

IntechOpen

chronically ill, particularly respiratory patients, constitute populations that are very sensitive to atmospheric pollution and are therefore usually chosen as a sample for studies in this area [11, 12]. It is known that children, a sensitive population, are more vulnerable to the effects of atmospheric pollution than adults, for several reasons, from the time they spend outdoors to the anatomy and physiology of the respiratory system, which is still under development. In addition, children have higher ventilation rates

	Unit	Variable description
Date		Observation date
Total	#	Total number of children observed
0–2	#	Number of children observed between 0 and 2 years old
3–5	#	Number of children observed between 3 and 5 years old
6–10	#	Number of children observed between 6 and 10 years old
11–15	#	Number of children observed between 11 and 15 years old
Cough	#	Number of children observed with cough symptoms
sdr	#	Number of children observed with of respiratory distress syndrome symptoms
Asthma	#	Number of children observed with asthma symptoms
Intern	#	Number of children admitted to hospital after observation
F	#	Number of female children observed
M	#	Number of male children observed
VEL (IM)	(km/h)	Average daily wind speed
DIREC		Prevailing daily wind direction
TEMP	(°C)	Average daily temperature
$TEMP_{MAX}$	(°C)	Maximum daily temperature
$TEMP_{MIN}$	(°C)	Minimum daily temperature
HUM	(%)	Humidity at 9 am
HUM_{max}	(%)	Humidity at 6 am
Rad	(watt/m^2)	Total daily radiation
$SO_{2\ med}$	(μg/m^3)	Average daily value of the average of all stations
NO_{med}	(μg/m^3)	Average daily value of the average of all stations
$NO_{2\ med}$	(μg/m^3)	Average daily value of the average of all stations
$NO_{x\ med}$	(μg/m^3)	Average daily value of the average of all stations
$PM_{10\ med}$	(ug/m^3)	Average daily value of the average of all stations
CO $(_{max})_{med}$	(μg/m^3)	Maximum daily value of eight-hour average by eight-hour averages.
O_3 $(_{max})_{med}$	(μg/m^3)	Maximum daily value of the eight-hour average
CO $(_{peak})_{med}$	(μg/m^3)	Maximum daily value, without considering eight-hour averages
O_3 $(_{peak})_{med}$	(μg/m^3)	Maximum daily value, without considering eight-hour averages
NO_x $(_{peak})_{med}$	(μg/m^3)	Maximum daily value

Table 1.
General description of the variables under study.

than adults and the short stature of children further increases their exposure to traffic emissions. All such factors contribute to the triggering of episodes of respiratory distress more frequently, even in the present of lesser pollutant concentrations [13].

The objective of this work was to study the relation between the number of children admitted to the emergency room (ER) of the pediatric service of Hospital Nª Srª do Rosário in Barreiro, with symptoms of respiratory problems, and the levels of atmospheric pollution, recorded by the air quality monitoring network, in the city of Barreiro.

2. Materials and methods

A team of pediatricians of Hospital Nossa Sra. do Rosário, in Barreiro, recorded daily, over a period of 20 months, the number and type of symptoms of children under the age of 15 years who were admitted to the hospital's pediatric emergency service. The children observed who had respiratory complaints of noninfectious etiology were classified according to their age, gender, area of residence, and type of symptom. As for symptoms, only three types of symptoms were considered: cough; respiratory distress (SDR), and asthma. Initially, the data collection by physicians was carried out with a frequency of two or three times a week. In a later phase of the project, collection became daily. As mentioned earlier, the tanking of symptoms by the medical team considered three groups only: cough, asthma, and breathing difficulty, although several international studies rank symptoms according to the List of International Statistical Classification of Diseases (ICD-11) [14], limiting the subjectivity inherent to the classification in these groups. In this case, a decision was made to simplify the classification due to the lack of a supporting computer system.

In parallel and during the same period, the values of pollutant concentrations from the Barreiro air quality network were used. The values of carbon monoxide (CO), nitrogen dioxide (NO_2), sulfur dioxide (SO_2), particles (PM_{10}), and ozone (O_3) were considered, as well as the meteorological variables of temperature and relative humidity. Meteorological variables, namely temperature, relative humidity, wind speed and direction, were provided by the Portuguese Institute of the Sea and the Atmosphere and/or by the CPPE of Barreiro. Thus, and during this period, the following data were recorded regarding the variables described in **Table 1**, for all observation stations.

A summary of the descriptive statistics on the study variables is shown in **Table 2**.

3. Results and discussion

To analyze and process the data, the statistical treatment programs R, SPSS, and Excel were used. The data analysis shows that during the period studied, 1101 children were assisted at the HNSR emergency service, with a daily average of 2.78 children, ranging from 0 to 12 children per day. One of the main challenges in understanding the effect on health of each of the pollutants and meteorological variables is related to the interdependence between these variables. Therefore, in a preliminary phase of this study, and in an attempt to understand the main relation between the most relevant variables, graphs representing two variables were generated, which allow verifying the joint evolution of these very variables.

FS	No. obs.	Ave.	Stand. devi.	Min. value	Max. value	Int.	Var. coef.	Median
Total	399	2.78	2.05	0	12	12	0.73	2
0–2	280	1.85	1.29	1	9	8	0.70	1
3–5	184	1.45	0.68	1	5	4	0.47	1
6–10	152	1.39	0.73	1	4	3	0.52	1
11–15	89	1.27	0.73	1	5	4	0.58	1
Cough	197	1.64	0.95	1	5	4	0.58	1
sdr	245	1.75	1.14	1	9	8	0.65	1
Asthma	197	1.63	0.91	1	5	4	0.56	1
Intern	38	1.11	0.31	1	2	1	0.28	1
F	247	1.71	0.99	1	6	5	0.58	1
M	301	2.14	1.34	0	9	9	0.63	2
VEL (IM)	730	4.87	6.58	0	40.5	40.5	1.35	3.1
DIREC	717	4.81	2.74	1	8	7	0.57	6
TEMP	721	17.72	4.48	7.7	31.0	23.3	0.25	17.4
$TEMP_{MAX}$	721	22.15	5.78	9.1	40.0	30.9	0.26	21.6
$TEMP_{MIN}$	721	14.16	4.06	0.0	26.0	26.0	0.29	14.1
HUM	723	71.93	18.26	6.8	99.1	92.3	0.25	68.0
HUM_{max}	711	84.71	12.58	39.3	99.1	59.8	0.15	87.6
Rad	153	5092.2	1778.02	697.0	7098.0	6401.0	0.35	5802.0
$SO_{2\ med}$	641	12.61	12.26	1.0	88.6	87.6	0.97	8.4
NO_{med}	641	13.98	15.39	1.1	136.1	135.1	1.10	9.1
$NO_{2\ med}$	641	29.98	11.94	5.8	72.5	66.7	0.40	28.5
$NO_{x\ med}$	641	48.94	27.92	9.1	188.9	179.8	0.57	42.8
$PM_{10\ med}$	617	38.04	21.88	3.5	152.5	148.9	0.58	31.7
$CO\ (_{max})_{med}$	641	491.52	367.15	167.6	2684.1	2516.5	0.75	362.8
$O_3\ (_{max})_{med}$	641	68.73	25.63	4.2	169.7	165.5	0.37	68.3
$CO\ (_{peak})_{ad}$	641	668.09	505.69	195.0	3740.5	3545.5	0.76	503.5
$O_3\ (_{peak})_{med}$	641	81.79	29.88	6.7	241.3	234.7	0.37	79.0
$NO_x\ (_{peak})_{med}$	641	88.78	54.91	9.0	452.8	443.8	0.62	79.4

Table 2.
Descriptive statistics of the variables collected.

In this way, **Figure 1** represents the average daily temperature evolution as a function of daily solar radiation. One can see that globally there is an increase of temperature with the increase of radiation. A Pearson correlation with the value $\rho = 0.46$ (95% CI: 0.33 to 0.58) is observed for these two variables.

Figure 2 represents the evolution of the daily maximum temperature as a function of daily solar radiation. One can see that globally there is an increase of the maximum temperature with the increase of radiation. A Pearson correlation is observed for these two variables with the value $\rho = 0.55$ (CI at 95%: 0.42 to 0.64).

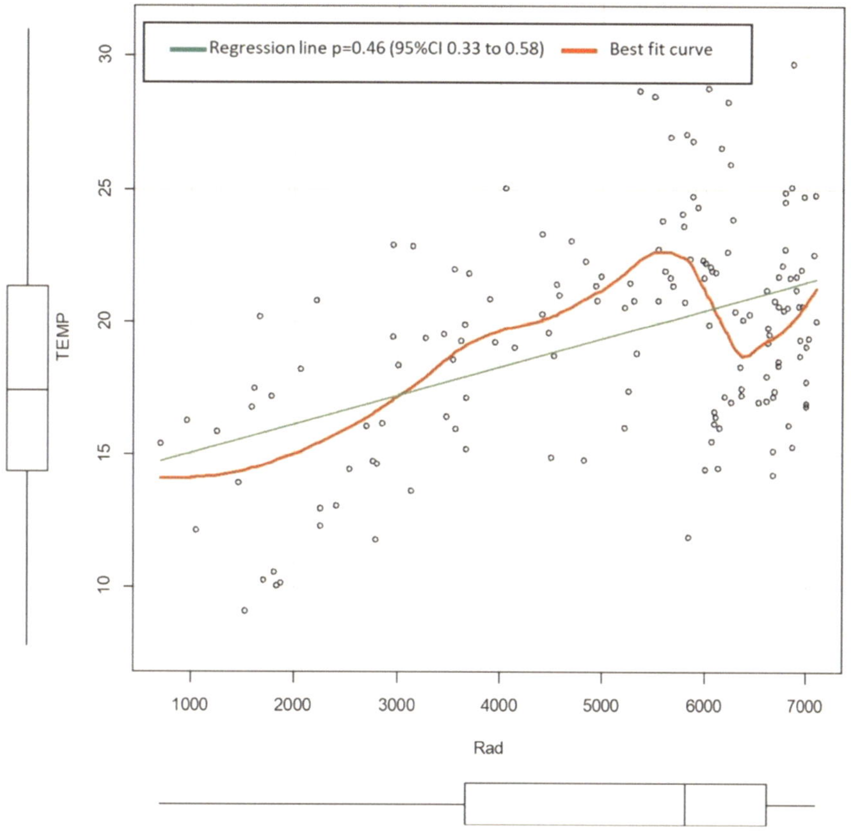

Figure 1.
Average temperature vs. radiation.

Figure 3 plots the evolution of the average daily value of PM_{10} concentration (average of all seasons) as a function of average daily temperature. One can see that, globally, there is an increase in the concentration of particles with the increase in temperature. However, looking at the evolution of the curve that best fits, one observes that for temperatures below 13°C there is a slight increase in particle concentration. This can be explained by the conditions of atmospheric stability that hinder the dispersion of particles. There is a Pearson correlation for these two variables with the value ρ = 0.22 (95% CI: 0.15 to 0.30), even though the values of average, maximum, and minimum daily temperatures show correlations above ρ = 0.9 between them. The other graphs with these temperatures show similar evolutions, as displayed in **Figure 4**, which presents the evolution of the average particle concentration with the daily maximum temperature. In this case, a Pearson correlation with the value ρ = 0.30 (95%CI 0.23 to 0.37) is seen for these two variables.

The detailed analysis of all data shows that the QA station that has the highest correlation between the average daily temperature and the PM_{10} concentration values (daily averages) is the QA station of Lavradio, so only the graph corresponding to this station is displayed (**Figure 5**).

For this specific station, the Pearson correlation coefficient between these variables has the value ρ = 0.44 (95% CI: 0.36 to 0.52).

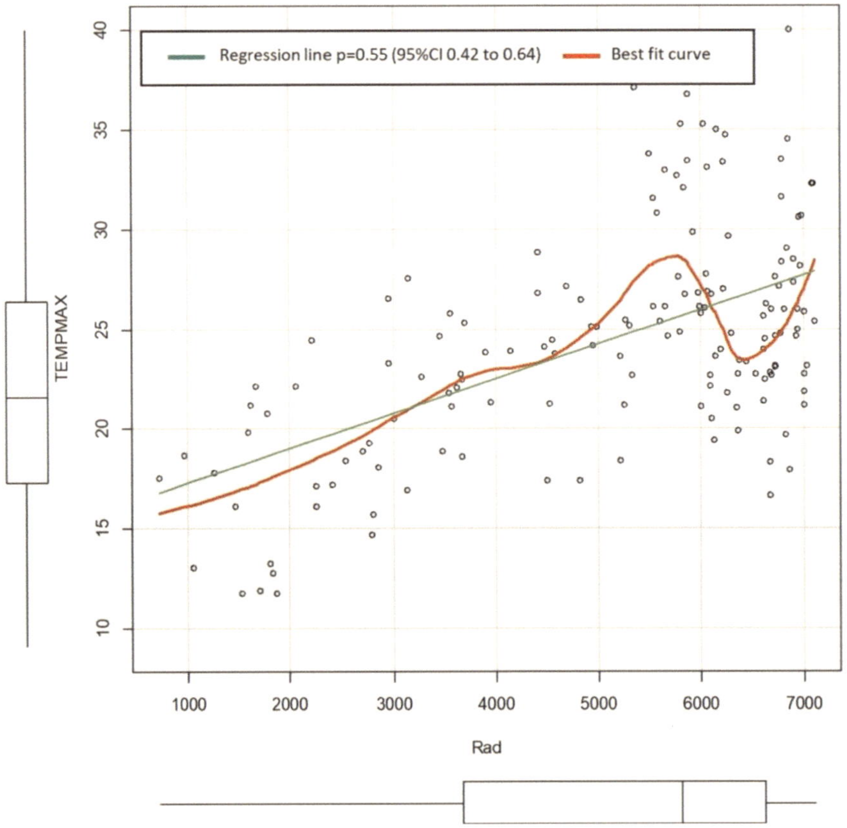

Figure 2.
Maximum temperature vs. radiation.

Figure 6 shows the evolution of the maximum ozone concentration (average of stations) as a function of average daily temperature. It appears that there is a direct relation between the concentration of ozone in the air and the average daily temperature, with a visible increase in concentration values with increasing temperature. This relation is indicated by the value of Pearson's correlation coefficient with the value $\rho = 0.53$ (95% CI: 0.47 to 0.58). This relation is even stronger for the QA station at the Hospital, where the correlation between ozone concentration and mean temperature has a Pearson correlation coefficient of $\rho = 0.57$ (95% CI: 0.52 to 0,63) as can be seen in **Figure** 7.

Figure 8 shows the values of the Pearson correlations (ρ) between the different variables collected at the hospital regarding the medical component. The analysis of **Figure 8** highlights important correlations between some variables analyzed. It appears that, for example, in the case of asthma, the correlation values increase with the increasing age of age groups. The correlation starts with a value of $\rho = 0.13$ for the 0–2 years age group, rising to $\rho = 0.26$ (3–5 years) and continuing to rise to $\rho = 0.39$ (6–10 years) reaching $\rho = 0.47$ (11–15 years), thus trending toward an increase in asthma symptoms, in visits to the hospital emergency room, with the increasing age of children.

Also noteworthy is the strong correlation between sdr symptoms and children aged 0 to 2 years ($\rho = 0.72$) and between cough symptoms and children aged 0 to 2 years ($\rho = 0.47$). There is also a strong correlation between hospital admission and children aged 6 to 10 years, indicating that older children only resort to the hospital emergency

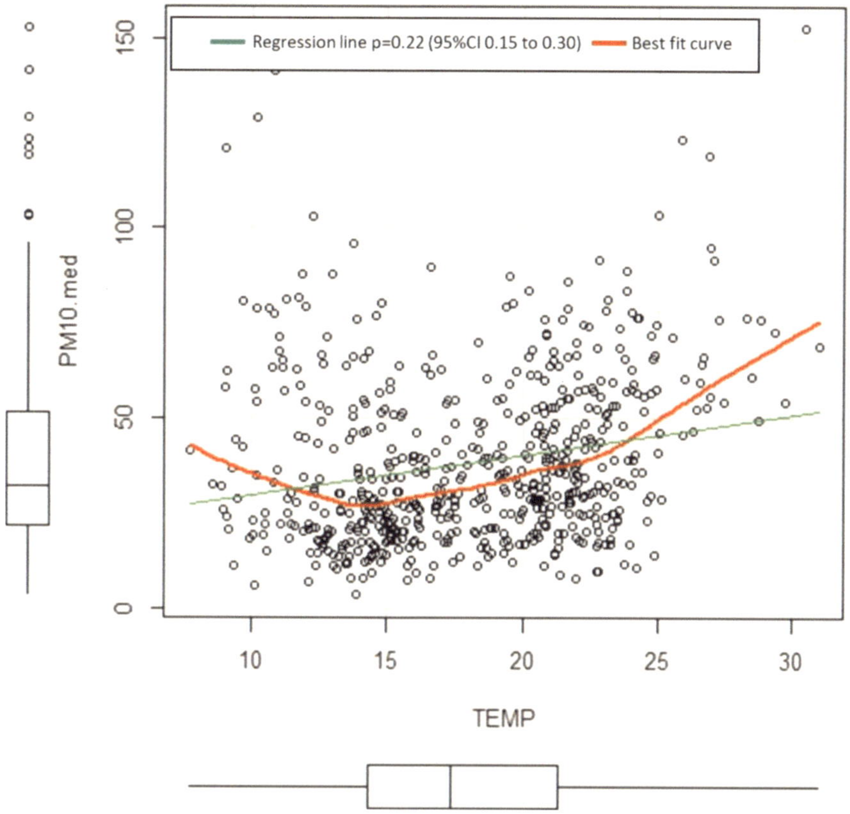

Figure 3.
Average concentration of particles vs. average temperature.

service in situations where seriousness justifies hospital admission, while younger children resort to the hospital emergency for all types of cases, including those that do not require hospital admission.

Figure 9 shows the values of the Pearson correlations (ρ) between the different variables related to the concentration of pollutants considering the mean values of all stations (global). Analyzing the values for Pearson's correlation coefficient (ρ), one can see that there are strong correlations between the average and peak values, for the same pollutant. The Pearson correlation coefficient values $\rho = 0.91$ (CO), $\rho = 0.95$ (O_3), and $\rho = 0.85$ (NO_x) (mean and peak) which expresses that, in fact, the values are very interdependent. This analysis allows concluding that the results obtained are not very different, using mean values of pollutant concentrations or peak values.

Carrying out a global assessment of the relation between pollutants, one can see that, in general, there are correlations between different pollutants, highlighting some important correlation values. For example, between the CO and NO_x families (NO, NO_2 and NO_x), the values of $\rho = 0.82$ (CO_{max} with NO_{med}), $\rho = 0.80$ (CO_{max} with $NO_{x\ med}$), $\rho = 0.84$ stand out (CO_{peak} with NO_{med}), and $\rho = 0.77$ (CO_{peak} with $NO_{x\ peak}$), indicating a strong relation between CO concentrations and NO_x families in an urban environment.

As for SO_2, it does not show strong correlations with any other pollutant, except for the value of the correlation coefficient of $\rho = 0.42$ between SO_2 and $O_{3\ peak}$ concentrations and $\rho = 0.36$ between SO_2 and $O_{3\ max}$ concentrations.

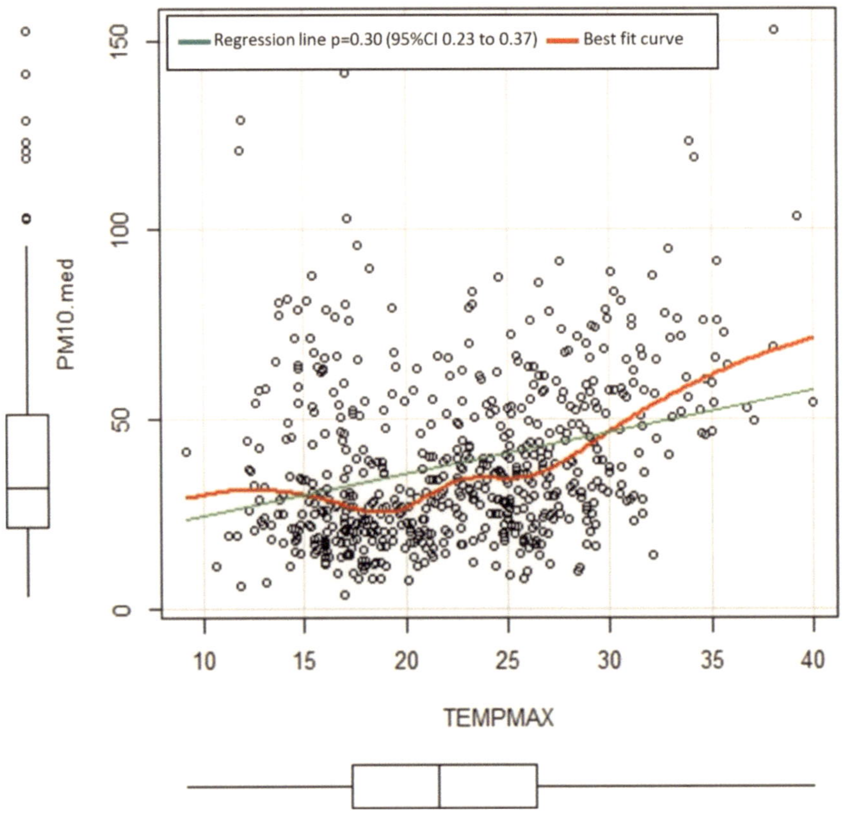

Figure 4.
Average concentration of particles vs. maximum temperature.

Regarding PM_{10}, the following important correlations are observed, ρ = 0.36 (PM_{10} with $NO_{x\ med}$), ρ = 0.47 (PM_{10} with $NO_{2\ med}$), ρ = 0.46 (PM_{10} with $NO_{x\ med}$), ρ =0.39 (PM_{10} with CO_{max}), ρ = 0.38 (PM_{10} with CO_{peak}), and ρ = 0.41 (PM_{10} with $NO_{x\ peak}$), indicating a relation between PM_{10} concentrations and CO concentrations and families of NO_x. As for the correlations with ozone, the values of ρ = 0.28 (PM_{10} with $O_{3\ peak}$) and ρ = 0.18 (PM_{10} with $O_{3\ max}$) suggest a lower correlation between these pollutants. Also, with regard to SO_2, the value of ρ = 0.29 (PM_{10} with $SO_{2\ med}$) indicates a lower correlation between PM_{10} concentrations and SO_2 concentrations.

One can thus see that PM_{10} concentrations are correlated with all pollutants, with the highest correlation for NO_2 and the lowest correlation for O_3. It should also be noted that ozone has negative correlations with NO_{med} (−0.58) and $NO_{x\ med}$ (−0.46), which can be explained by the fact that NO_x are precursors of Ozone, causing the formation of each molecule of O_3 to lead to a consequent decrease in NO_x.

Figure 10 shows the values of the Pearson correlations (ρ) between all variables analyzed in this study (medical and pollutant variables). Being an extensive figure, the strong correlations between the average concentration values for the different pollutants and the concentration values of said pollutants for the different air quality stations stand out.

Figure 11 shows the Pearson correlation values (ρ) between the medical variables studied and the mean concentrations of pollutants measured at the stations. This figure is important because it allows analyzing the possible relation between the concentrations of pollutants and the symptoms studied. Analyzing the figure one can

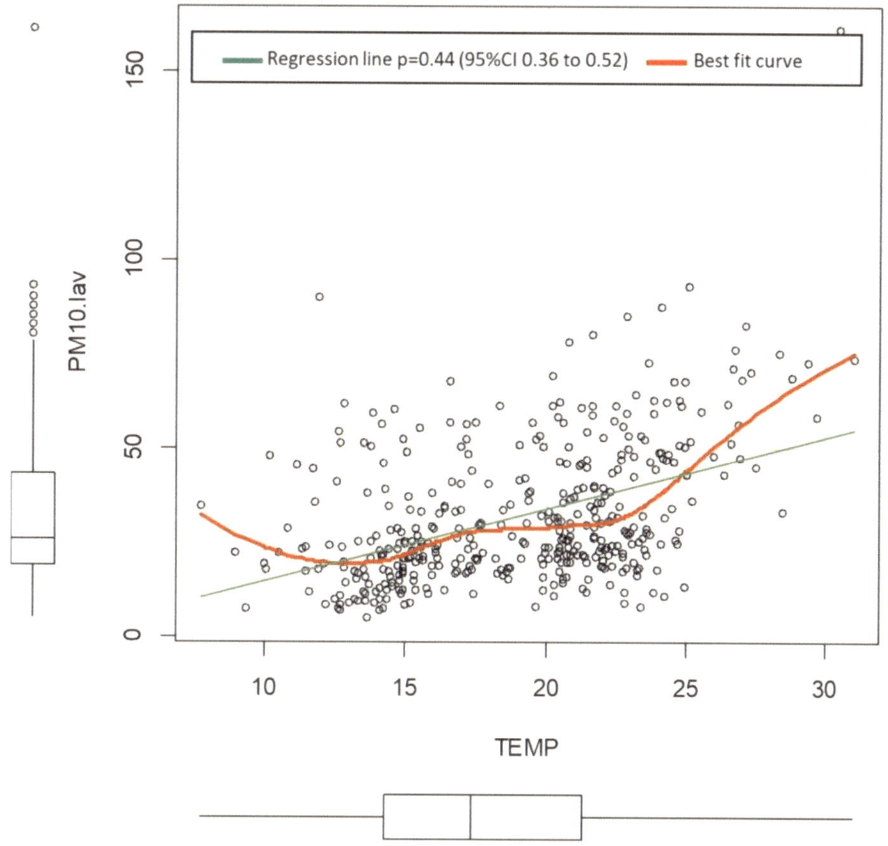

Figure 5.
Particle concentration (Lavradio station) vs. temperature.

see, globally, that the more important values of correlation between the pollutants and the symptoms are obtained for the symptomatology of asthma, with the highest values for the correlation coefficient with CO_{max} $\rho = 0.26$ and CO_{peak} $\rho = 0.25$. There is also an important correlation between asthma and NO_x, $\rho = 0.21$ (with $NO_{x\ med}$), $\rho = 0.21$ (with $NO_{x\ med}$) $\rho = 0.21$ (with $NO_{x\ peak}$). The correlation with PM_{10}, however, has an unrepresentative value of $\rho = 0.09$.

Negative correlation values are also seen between asthma symptoms and O_3 $\rho = -0.23$ (asthma and $O_{3\ max}$) and $\rho = -0.22$ (asthma and $O_{3\ peak}$), as well as a negative correlation between asthma and $SO_{2\ med}$ ($\rho = -,12$).

Figure 12 shows the values of the Pearson correlations (ρ) between all variables referring to the atmospheric concentrations of the pollutants studied. An analysis of this figure reveals, as expected, the existence of important correlations between the average values of the concentrations of the different pollutants and the values of the concentrations of these same pollutants, in the different air quality stations.

With the aim of studying the evolution of the number of children observed in hospital with the symptomatology of asthma, sdr, or cough, it was analyzed the number of days elapsed, considering temporal lags of 0, 1, 2, 3, 4, 5, 6, 7, and 15 days between the values of PM_{10} concentrations and the aforementioned medical variables.

Table 3 and **Figure 13** show the evolution of the Pearson correlation coefficient (ρ) between the asthma variables and the average concentrations of pollutants (O_3, SO_2,

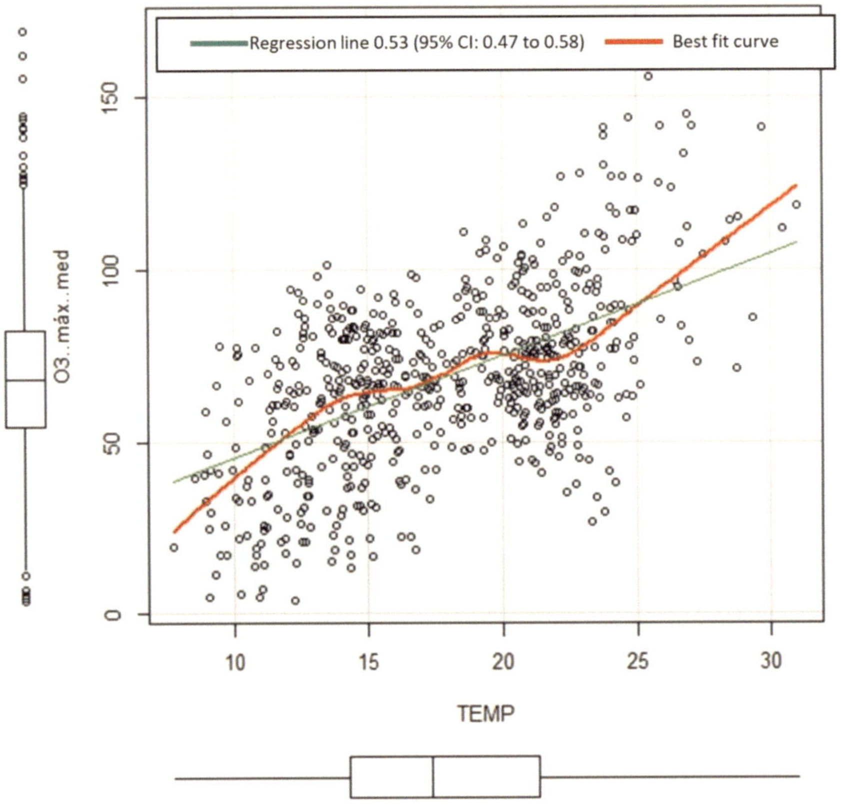

Figure 6.
Maximum ozone concentration (average of stations) vs. average air temperature.

NO, NO_2, NO_x PM_{10}, and CO) for the temporal lags of 0, 1, 2, 3, 4, 5, 6, 7, and 15 days. It appears that, overall, the maximum correlations between asthma symptoms and the concentrations of pollutants NO, NO_2, NO_x, CO, and PM_{10} occur after lags of 2, 3, or 4 days (depending on the symptoms and the pollutant analyzed). The following Pearson's correlation coefficient values are also observed, ρmax = 0.35 (asthma and CO, 2-day lag), ρmax = 0.32 (asthma and NO, 3-day lag), ρmax = 0.33 (asthma and NO_x, lag 3 days), ρmax = 0.22 (asthma and NO_2, lag 2 days), and ρmax = 0.18 (asthma and PM_{10}, lag 4 days). There is, for these pollutants, an increase in Pearson's correlation coefficient values (ρ), over the days until it reaches its maximum after 2, 3, or 4 days. For PM_{10} and NO_2 there is a slight decrease in the correlation coefficient from lag 0 to lag 1, this coefficient increasing again from lag 2 onward.

For SO_2 and O_3 pollutants, it appears that Pearson's correlation coefficient (ρ) remains relatively constant and negative throughout the period analyzed, (ρ approx −0.2; asthma and O_3) and (ρ approx −0,1; asthma and SO_2) for all lags analyzed, suggesting lack of short-term correlations between observations with asthma symptoms and atmospheric concentrations of these two pollutants.

Table 4 and **Figure 14** show the evolution of the Pearson correlation coefficient (ρ) between the variable sdr and the mean concentrations of pollutants (O_3, SO_2, NO, NO_2, NO_x, PM_{10}, and CO) for the temporal lags of 0, 1, 2, 3, 4, 5, 6, 7, and 15 days. One can see that, globally, the maximum correlations between the symptomatology of

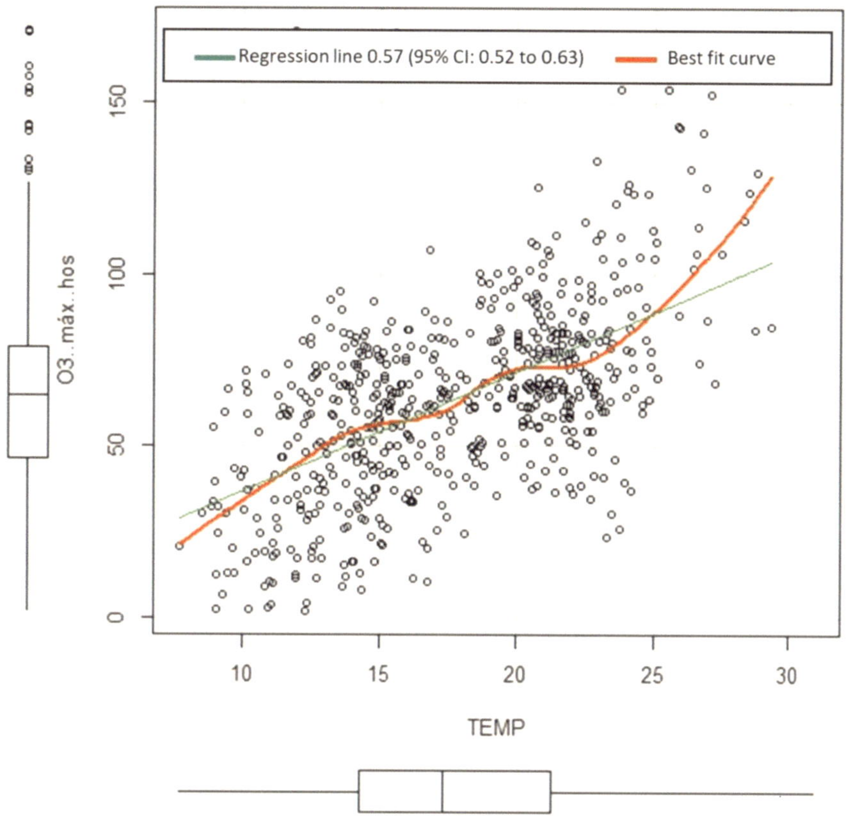

Figure 7.
Maximum ozone concentration (hospital QA station) vs. average air temperature.

			N.C				Doenças			Sexo	
	Total	0-2	3-5	6-10	11-15	tosse	sdr	asma	intern	F	M
Total	1.00										
0-2	0.76	1.00									
3-5	0.35	0.00	1.00								
6-10	0.37	0.14	0.02	1.00							
11-15	0.38	0.24	0.12	-0.22	1.00						
tosse	0.55	0.47	0.26	0.03	0.22	1.00					
sdr	0.72	0.72	0.15	0.27	0.12	0.19	1.00				
asma	0.55	0.13	0.26	0.39	0.47	0.03	0.20	1.00			
intern	0.25	0.25	0.13	0.77	-0.23	-0.17	0.37	0.04	1.00		
F	0.72	0.55	0.31	0.20	0.10	0.50	0.51	0.31	-0.03	1.00	
M	0.81	0.66	0.14	0.28	0.43	0.28	0.65	0.46	0.29	0.25	1.00

Figure 8.
Medical data correlations.

sdr and the concentrations of pollutants NO, NO_2, NO_x, CO, and PM_{10} occur after temporal lags of 4 days. Observing the following Pearson correlation coefficient values, ρmax = 0.20 (sdr and NO_x, 4 day lag), ρmax = 0.17 (sdr and NO_2, 4 day lag), ρmax = 0.15 (sdr and PM_{10}, lag 4 days), ρmax = 0.14 (sdr and CO, lag 4 days), ρmax = 0.13 (sdr and NO, lag 4 days). For these pollutants (NO, NO_2, NO_x, PM_{10}, and CO), there is a decrease in the correlation coefficient from lag 0 to lag 1, and this coefficient increases again from lag 2 and up to lag 4.

For SO_2 and O_3 pollutants, one can see that Pearson's correlation coefficient (ρ) remains relatively constant and negative throughout the period analyzed, for all lags

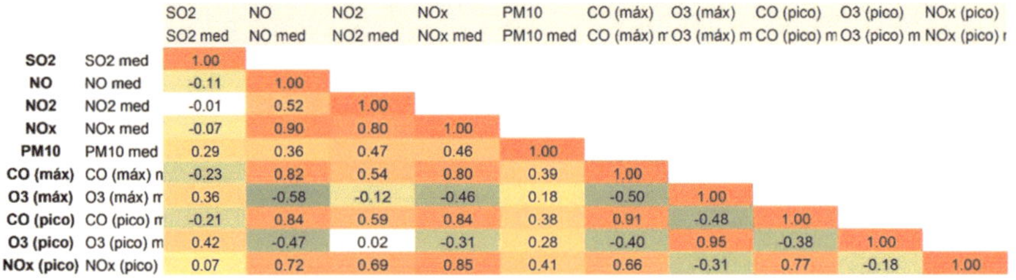

		SO2	NO	NO2	NOx	PM10	CO (máx)	O3 (máx)	CO (pico)	O3 (pico)	NOx (pico)
		SO2 med	NO med	NO2 med	NOx med	PM10 med	CO (máx) m	O3 (máx) m	CO (pico) m	O3 (pico) m	NOx (pico) r
SO2	SO2 med	1.00									
NO	NO med	-0.11	1.00								
NO2	NO2 med	-0.01	0.52	1.00							
NOx	NOx med	-0.07	0.90	0.80	1.00						
PM10	PM10 med	0.29	0.36	0.47	0.46	1.00					
CO (máx)	CO (máx) n	-0.23	0.82	0.54	0.80	0.39	1.00				
O3 (máx)	O3 (máx) m	0.36	-0.58	-0.12	-0.46	0.18	-0.50	1.00			
CO (pico)	CO (pico) m	-0.21	0.84	0.59	0.84	0.38	0.91	-0.48	1.00		
O3 (pico)	O3 (pico) m	0.42	-0.47	0.02	-0.31	0.28	-0.40	0.95	-0.38	1.00	
NOx (pico)	NOx (pico)	0.07	0.72	0.69	0.85	0.41	0.66	-0.31	0.77	-0.18	1.00

Figure 9.
Correlations between pollutants.

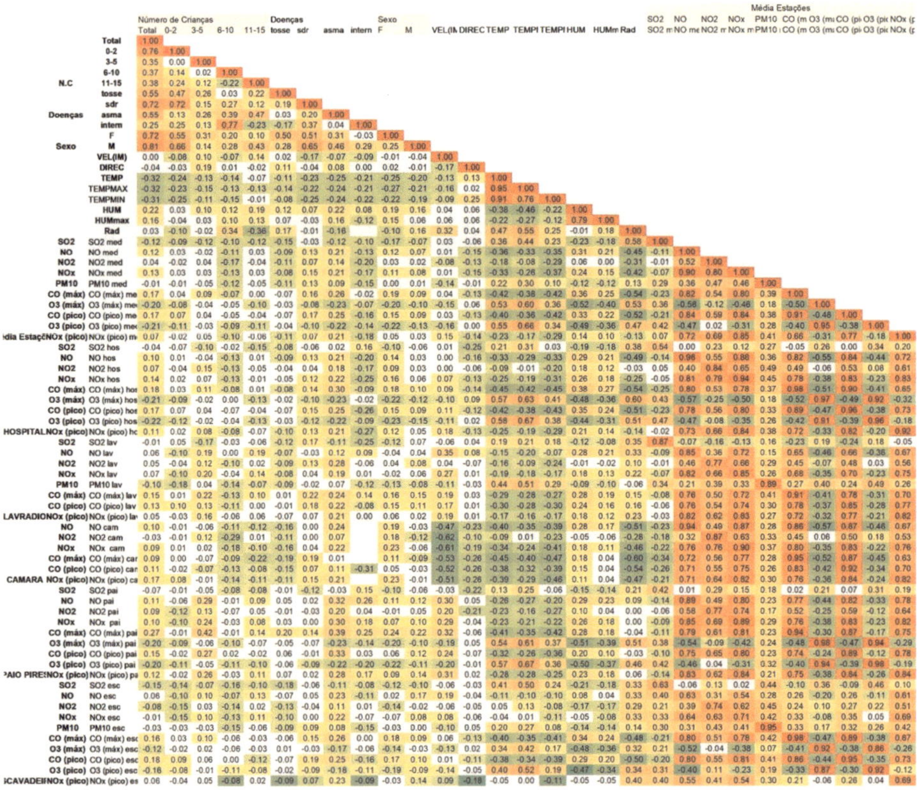

Figure 10.
Correlations between all variables studied.

considered, suggesting the inexistence of short-term correlations between observations with sdr symptoms and atmospheric concentrations of these two pollutants.

Table 5 and **Figure 15** show the evolution of the Pearson correlation coefficient (ρ) between the cough variable and the mean concentrations of pollutants (O_3, SO_2, NO, NO_2, NO_x, PM_{10}, and CO) for the temporal lags of 0, 1, 2, 3, 4, 5, 6, 7, and 15 days.

One can see that, globally, the maximum correlations between the symptoms of cough and the concentrations of pollutants NO, NO_2, NO_x, CO, and PM_{10} occur after temporal lags of 6 days, except for SO_2, which has the maximum correlation with a temporal lag of 5 days, albeit with a maximum correlation value of ρmax = 0.07 (cough and SO_2, lag 5 days), which is not significant. In addition, the following Pearson correlation coefficient values stand out, ρmax = 0.20 (cough and PM_{10}, 6-day

	lag 0	lag 1	lag 2	lag 3	lag 4	lag 5	lag 6	lag 7	lag 15
O_3	−0.23	−0.19	−0.26	−0.26	−0.26	−0.27	−0.26	−0.25	−0.06
SO_2	−0.12	−0.09	−0.10	−0.08	−0.09	−0.05	−0.08	−0.14	−0.06
NO	0.21	0.23	0.30	0.32	0.26	0.25	0.31	0.20	−0.02
NO_2	0.14	0.09	0.22	0.21	0.22	0.18	0.15	0.12	0.01
NO_x	0.21	0.20	0.30	0.33	0.28	0.25	0.30	0.21	0.00
PM_{10}	0.09	0.00	0.06	0.12	0.18	0.11	0.09	0.06	−0.07
CO	0.26	0.28	0.35	0.33	0.27	0.34	0.29	0.26	0.12

Table 3.
Pearson's correlation coefficient (ρ) between the variable mean concentrations of pollutants (O_3, SO_2, NO, NO_2, NO_x, PM_{10}, and CO) and asthma for time lags of 0, 1, 2, 3, 4, 5, 6, 7, and 15 days.

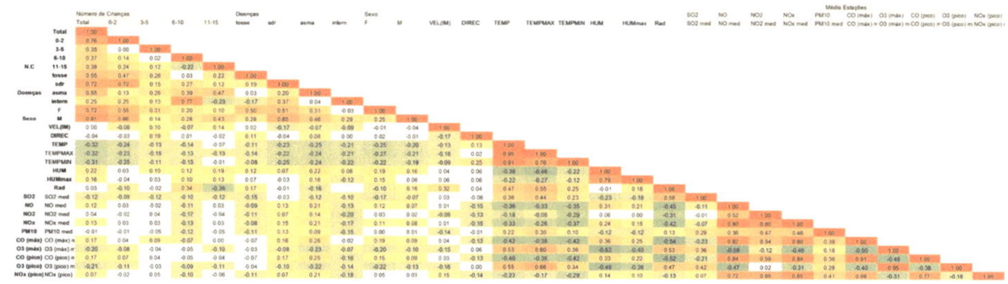

Figure 11.
Correlations of medical data with station averages.

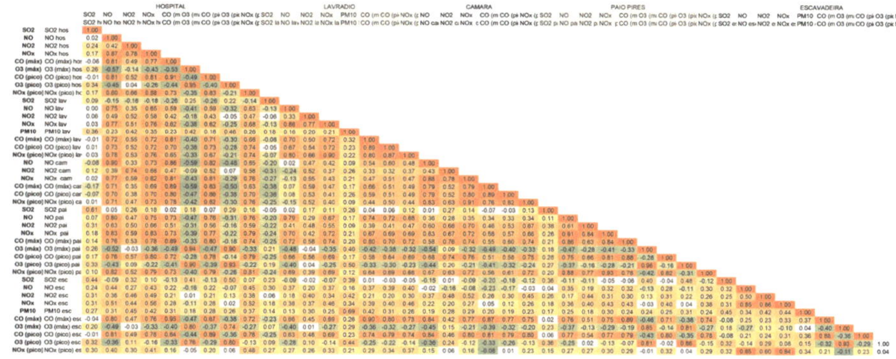

Figure 12.
Correlations between seasons.

lag), ρmax = 0.20 (cough and NO_x, 6-day lag), ρmax = 0, 18 (cough and NO, lag 6 days), ρmax = 0.17 (cough and CO, lag 6 days), and ρmax = 0.09 (cough and NO_2, lag 6 days), representing nonsignificant values.

For O_3, it appears that Pearson's correlation coefficient (ρ) remains relatively constant and negative throughout the period analyzed for all lags considered, indicating the inexistence of short-term correlations between observations with symptoms of cough and atmospheric concentrations of this pollutant.

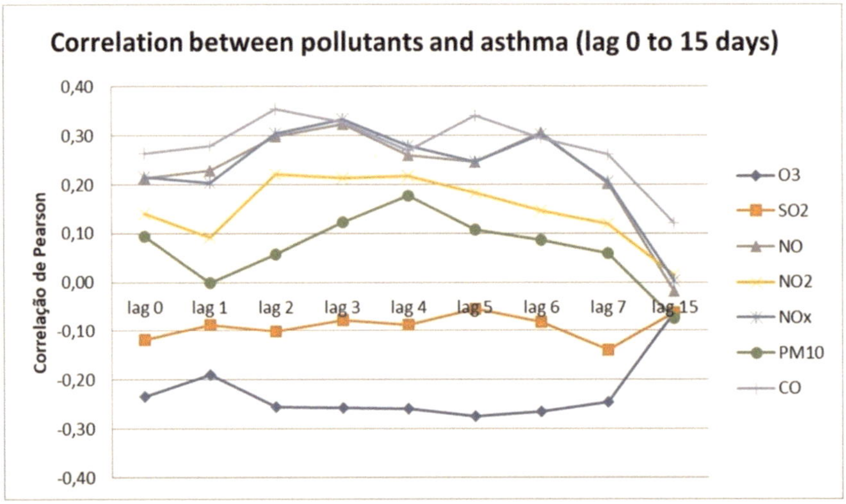

Figure 13.
Evolution of the correlation coefficient between concentrations of air pollutants and asthma symptoms (lag 0 to 15 days).

	lag 0	lag 1	lag 2	lag 3	lag 4	lag 5	lag 6	lag 7	lag 15
O_3	−0.08	−0.11	−0.11	−0.16	−0.15	−0.14	−0.14	−0.13	−0.17
SO_2	−0.03	−0.04	−0.04	−0.11	0.01	−0.05	−0.17	−0.11	−0.09
NO	0.13	0.09	0.11	0.12	0.13	0.07	0.13	0.11	0.12
NO_2	0.07	0.01	0.07	0.12	0.17	0.13	0.14	0.09	−0.02
NO_x	0.15	0.09	0.13	0.16	0.20	0.14	0.19	0.15	0.10
PM_{10}	0.13	0.00	0.07	0.14	0.15	0.12	0.05	0.08	−0.03
CO	0.16	0.08	0.14	0.11	0.14	0.09	0.12	0.10	0.15

Table 4.
Pearson's correlation coefficient (ρ) between the variable mean concentrations of pollutants (O_3, SO_2, NO, NO_2, NO_x, PM_{10}, and CO) and sdr for temporal lags of 0, 1, 2, 3, 4, 5, 6, 7, and 15 days.

4. Conclusions

The results presented in this chapter allow concluding some important aspects for the study and understanding the relation between the concentrations of atmospheric pollutants, meteorological variables, such as air temperature and solar radiation, and medical variables related to the different symptomatologies studied, in admissions of children in the ER, at Hospital do Barreiro (sdr, cough, and asthma). From the development and analysis of the results of this study, the following aspects stand out:

There is an increase of the ambient air temperature with the increase of solar radiation. This fact is visible in the values of Pearson's correlation coefficient $\rho = 0.46$ (mean daily air temperature with radiation) and $\rho = 0.55$ (maximum daily air temperature with radiation).

There is an increase in the concentration of PM_{10} with the increase of the ambient air temperature. This is more relevant for temperatures above 25°C - $\rho = 0.22$ (average

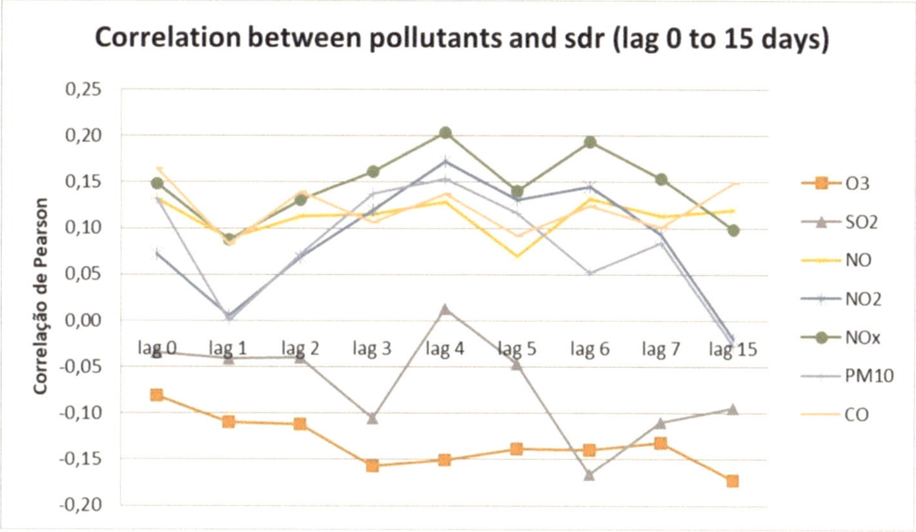

Figure 14.
Evolution of the correlation coefficient between air pollutant concentrations and sdr symptomatology (lag 0 to 15 days).

	lag 0	lag 1	lag 2	lag 3	lag 4	lag 5	lag 6	lag 7	lag 15
O_3	−0.03	−0.09	−0.11	−0.15	−0.16	−0.19	−0.17	−0.12	−0.14
SO_2	−0.15	−0.11	−0.05	−0.02	−0.07	0.07	0.01	−0.02	−0.07
NO	−0.09	−0.02	0.03	0.03	0.06	0.15	0.18	0.06	0.10
NO_2	−0.11	−0.01	0.02	0.02	0.05	0.02	0.09	−0.02	−0.05
NO_x	−0.08	0.01	0.04	0.06	0.10	0.15	0.20	0.08	0.07
PM_{10}	−0.11	−0.07	0.03	−0.03	−0.04	0.11	0.20	0.11	0.02
CO	−0.07	−0.03	−0.03	0.02	0.09	0.12	0.17	0.12	0.07

Table 5.
Pearson's correlation coefficient (ρ) between the variable mean concentrations of pollutants (O_3, SO_2, NO, NO_2, NO_x, PM_{10}, and CO) and cough for temporal lags of 0, 1, 2, 3, 4, 5, 6, 7, and 15 days.

concentration of PM_{10} with average daily air temperature) and ρ = 0.30 (average concentration of PM_{10} with maximum daily air temperature). One can also see that for air temperatures below 13°C there is a slight increase in the concentration of particles, probably due to atmospheric conditions that hinder the dispersion of particles.

There is a strong correlation between the maximum concentration of ozone in the air and the average daily air temperature ρ = 0.53, with an increase in the values of ozone concentration with the increase in average air temperature.

There is a trend toward an increase in asthma symptoms, in visits to the hospital's emergency room, with increasing age of children (ρ = 0.13 0–2 years, ρ = 0.26 3–5 years; ρ = 0.39 6–10 years; ρ = 0.47 11–15 years).

There are important correlations between the values of the concentrations of the different atmospheric pollutants, correlated among themselves, there being a strong

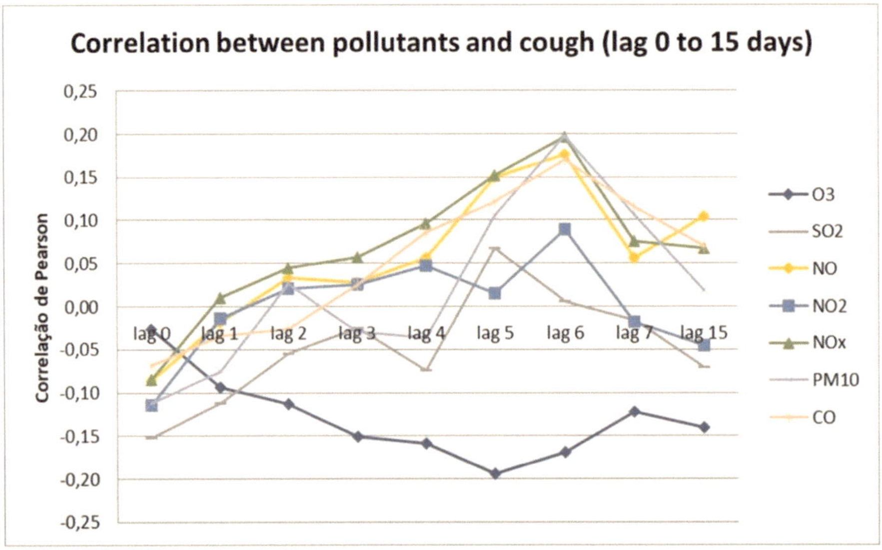

Figure 15.
Evolution of the correlation coefficient between concentrations of air pollutants and cough symptomatology (lag 0 to 15 days).

relation between the concentrations of CO and the NOx families (ρ = 0.82 CO_{max} with NO_{med}, ρ = 0.80 CO_{max} with $NO_{x\ med}$, ρ = 0.84 CO_{peak} with NO_{med}, ρ = 0.77 CO_{peak} with $NO_{x\ peak}$). On the contrary, SO_2 does not show strong correlations with any other pollutant, except for an important correlation value between peak SO_2 and O_3 concentrations (ρ = 0.42) and between SO_2 and $O_{3\ max}$ concentrations (ρ = 0,36). PM_{10} concentrations are correlated with the concentrations of all pollutants, with the highest correlation for NO_2 (ρ = 0.47 PM_{10} with $NO_{2\ med}$) and the lowest correlation for O_3 (ρ = 0.28 PM_{10} with $O_{3\ peak}$; ρ = 0.18 PM_{10} with $O_{3\ max}$).

The strongest correlations between the variables of health symptoms (sdr, cough, and asthma) and the concentrations of air pollutants for situations of lag 0 (consequences on the same day) are observed for the symptoms of asthma, presenting the highest values for the correlation coefficient with CO_{max} (ρ = 0.26 asthma CO_{max}) and with CO_{peak} (ρ = 0.25 asthma CO_{peak}). There is also an important correlation between asthma and NO_x for situations of lag 0, ρ = 0.21 (asthma with $NO_{x\ med}$), ρ = 0.21 (asthma with $NO_{x\ med}$), and ρ = 0.21 (asthma with peak NO_x). The correlation with PM_{10}, on the other hand, has an unrepresentative value ρ = 0.09 (asthma with PM_{10}) for the situation of lag 0 (same day);

For the situation of temporal lag 0 analyzed (consequences on the same day), negative correlation values are observed between asthma symptoms and O_3 $\rho = -0.23$ (asthma and $O_{3\ max}$) and $\rho = -0.22$ (asthma and peak O_3), as well as a negative correlation between asthma and $SO_{2\ med}$ ($\rho = -0.12$);

Staggering the analysis among health symptoms (sdr, cough, and asthma) and the concentrations of air pollutants (O_3, SO_2, NO, NO_2, NO_x, PM_{10}, and CO) considering temporal lags of 1, 2, 3, 4, 5, 6, 7, 8, and 15 days, overall, the maximum correlations between these symptoms and the concentrations of pollutants NO, NO_2, NO_x, CO, and PM_{10} occur after time lags ranging from 2 to 6 days depending on the pollutant and the symptomatology considered;

For SO_2 and O_3 pollutants, analyzing the evolution of the Pearson correlation coefficient (ρ) between these pollutants and the symptoms analyzed, for temporal lags of 1, 2, 3, 4, 5, 6, 7, 8, and 15 days, this correlation coefficient remains relatively constant or negative throughout the period analyzed, indicating lack of short-term correlations between observations with asthma symptoms and concentrations of atmospheric emissions of these two pollutants.

Author details

João Garcia[1,2]*, Rita Cerdeira[3] and Luís Coelho[2,4]

1 Lisbon Superior Institute of Engineering, Lisbon, Portugal

2 CINEA-IPS, Energy and Environment Research Centre, IPS Campus, Setúbal, Portugal

3 Independent Researcher, Setúbal, Portugal

4 Mechanical Engineering Department, Polytechnic Institute of Setúbal, Higher School of Technology, Setúbal, Portugal

*Address all correspondence to: joao.garcia@isel.pt

IntechOpen

© 2023 The Author(s). Licensee IntechOpen. This chapter is distributed under the terms of the Creative Commons Attribution License (http://creativecommons.org/licenses/by/3.0), which permits unrestricted use, distribution, and reproduction in any medium, provided the original work is properly cited.

References

[1] HEI. State of Global Air 2018. Special Report. Boston, MA: Health Effects Institute; 2018

[2] European Environment Agency. Air quality in Europe 2022; Technical Report No 05/2022. 2022. DOI: 10.2800/488115

[3] Vallius M. Characteristics and Sources of Fine Particulate Matter in Urban Air. Kuopio, Finland: National Public Health Institute, Department of Environmental Health; 2005

[4] Garcia J, Cerdeira R, Coelho L, Kumar P, Carvalho MG. Important variables when studying the influence of particulate matter on health. In: Particulate Matter: Sources, Impact and Health Effects. Setubal, Portugal: Nova Publishers; 2017

[5] Wang X-Q, Huang K, Cheng X, Cheng-Yang H, Ding K, Yang X-J, et al. Short-term effect of particulate air pollutant on the risk of tuberculosis outpatient visits: A multicity ecological study in Anhui, China. Atmospheric Environment. 2022;**280**:119129. DOI: 10.1016/j.atmosenv.2022.119129. ISSN 1352-2310

[6] Usmani RSA, Pillai TR, Hashem IAT, et al. Air pollution and cardiorespiratory hospitalization, predictive modeling, and analysis using artificial intelligence techniques. Environmental Science and Pollution Research. 2021;**28**: 56759-56771. DOI: 10.1007/s11356-021-14305-7

[7] Guilbert A, Cox B, Bruffaerts N, et al. Relationships between aeroallergen levels and hospital admissions for asthma in the Brussels-capital region: A daily time series analysis. Environmental Health. 2018;**17**:35. DOI: 10.1186/s12940-018-0378-x

[8] Nel A. Air pollution-related illness: Effects of particles. Science. 2015; **308**(5723):804-806. DOI: 10.1126/science.1108752

[9] European Environment Agency. Air Quality in Europe – 2011 Report. EEA Technical Report. Denmark – Copenhagen; 2011. no. 12/2011

[10] Gasparrini A. Modeling exposure-lag-response associations with distributed lag non-linear models. Statistics in Medicine. 2014;**33**(5): 881-899. DOI: 10.1002/sim.5963. Available from: wileyonlinelibrary.com

[11] Manisalidis I, Stavropoulou E, Stavropoulos A, Bezirtzoglou E. Environmental and health impacts of air pollution: A review. Frontiers in Public Health. 2020;**8**:14. DOI: 10.3389/fpubh.2020.00014

[12] Dhital S, Rupakheti D. Correction to: Bibliometric analysis of global research on air pollution and human health: 1998–2017. Environmental Science and Pollution Research International. 2019 Aug;**26**(24):25386. DOI: 10.1007/s11356-019-05792-w. Erratum for: Environmental Science and Pollution Research International. 2019 May;26 (13):13103-13114. PMID: 31240646

[13] Sicard P, Agathokleous E, De Marco A, et al. Urban population exposure to air pollution in Europe over the last decades. Environmental Sciences Europe. 2021;**33**:28. DOI: 10.1186/s12302-020-00450-2

[14] World Health Organization. ICD-11 Implementation or Transition Guide. Geneva: License: CC BY-NC-SA 3.0 IGO; 2019